WITHDRAWN
UTSA LIBRARIES

China on the Move

The 150 million migrants that left home in pursuit of economic opportunities are changing the face of China. This book is about this human tide, how it has evolved over the last two decades, and what it means to China's city and countryside. By addressing the hukou paradigm and the permanent migrant paradigm, the book challenges the assumption that peasant migrants' main goal is permanent residence in the city. Rather, the findings show that splitting the peasant household into a permanent rural segment and a temporary urban segment allows migrants to straddle and benefit from two worlds. Circulation between the city and the countryside and among places of migrant work, rather than permanent migration, is becoming a way of life and an essential means of economic betterment for Chinese peasants. This strategy entails negotiation within and across households and changes in gender division of labor.

This book combines a macro structural perspective that focuses on the post-Mao state and a micro bottom-up approach that highlights household strategies to interpret the massive population movement in China. Based on detailed analysis of census and survey data and insights from field materials and migrants' narratives, the book documents and identifies key forces that shape migrants' decision-making. It also examines changes in the composition and spatial pattern of migration, the hukou (household registration) system and its evolution, differentials between male and female migration, gender relations and power hierarchy, circular migration, migrants' experiences in the city, the impacts of migration on the countryside, marriage migration, and new opportunities and challenges for the Chinese on the move in the twenty-first century.

The book draws on twenty years of English-language and Chinese-language studies on migration, and it depicts salient empirical information via maps, illustrations and tables. Population researchers, Asian Studies scholars, China observers and specialists, and students ranging from advanced undergraduates to doctoral students will find this book both informative and stimulating.

C. Cindy Fan is Professor at the UCLA Geography Department. Her research examines regional and social questions in transitional economies, focusing on labor migration, gender, regional inequality, and population in post-Mao China. She has published more than fifty refereed research articles and is editor of *Regional Studies* and senior contributing editor of *Eurasian Geography and Economics*.

Routledge studies in human geography

This series provides a forum for innovative, vibrant, and critical debate within Human Geography. Titles will reflect the wealth of research which is taking place in this diverse and ever-expanding field.

Contributions will be drawn from the main sub-disciplines and from innovative areas of work which have no particular sub-disciplinary allegiances.

Published:

1 **A Geography of Islands**
   Small island insularity
   Stephen A. Royle

2 **Citizenships, Contingency and the Countryside**
   Rights, Culture, Land and the Environment
   Gavin Parker

3 **The Differentiated Countryside**
   Jonathan Murdoch, Philip Lowe, Neil Ward and Terry Marsden

4 **The Human Geography of East Central Europe**
   David Turnock

5 **Imagined Regional Communities**
   Integration and sovereignty in the global south
   James D Sidaway

6 **Mapping Modernities**
   Geographies of Central and Eastern Europe 1920–2000
   Alan Dingsdale

7 **Rural Poverty**
   Marginalisation and exclusion in Britain and the United States
   Paul Milbourne

8 **Poverty and the Third Way**
  Colin C. Williams and Jan Windebank

9 **Ageing and Place**
  Edited by Gavin J. Andrews and David R. Phillips

10 **Geographies of Commodity Chains**
   Edited by Alex Hughes and Suzanne Reimer

11 **Queering Tourism**
   Paradoxical performances at Gay Pride Parades
   Lynda T. Johnston

12 **Cross-Continental Food Chains**
   Edited by Niels Fold and Bill Pritchard

13 **Private Cities**
   Edited by Georg Glasze, Chris Webster and Klaus Frantz

14 **Global Geographies of Post Socialist Transition**
   Tassilo Herrschel

15 **Urban Development in Post-Reform China**
   Fulong Wu, Jiang Xu and Anthony Gar-On Yeh

16 **Rural Governance**
   International perspectives
   Edited by Lynda Cheshire, Vaughan Higgins and Geoffrey Lawrence

17 **Global Perspectives on Rural Childhood and Youth**
   Young rural lives
   Edited by Ruth Panelli, Samantha Punch, and Elsbeth Robson

18 **World City Syndrome**
   Neoliberalism and inequality in Cape Town
   David A. McDonald

19 **Exploring Post Development**
   Aram Ziai

20 **Family Farms**
   Harold Brookfield and Helen Parsons

21 **China on the Move**
   Migration, the state, and the household
   C. Cindy Fan

Not yet published:

22 Participatory Action Research Approaches and Methods
Connecting people, participation and place
Sara Kindon, Rachel Pain and Mike Kesby

23 Historical Geographies of Time Space Compression
Barney Warf

24 International Migration and Knowledge
Allan Williams and Vladimir Balaz

China on the Move
Migration, the state, and the household

C. Cindy Fan

LONDON AND NEW YORK

First published 2008
by Routledge
2 Park Square, Milton Park, Abingdon, Oxon OX14 4RN

Simultaneously published in the USA and Canada
by Routledge
270 Madison Ave, New York, NY 10016

Routledge is an imprint of the Taylor & Francis Group, an informa business

© 2008 C. Cindy Fan

Typeset in Times by Wearset Ltd, Boldon, Tyne and Wear
Printed and bound in Great Britain by MPG Books Ltd, Bodmin

All rights reserved. No part of this book may be reprinted or reproduced or utilized in any form or by any electronic, mechanical, or other means, now known or hereafter invented, including photocopying and recording, or in any information storage or retrieval system, without permission in writing from the publishers.

British Library Cataloguing in Publication Data
A catalogue record for this book is available from the British Library

Library of Congress Cataloging in Publication Data
Fan C. Cindy, 1960–
China on the move: migration, the state, and the household / by C. Cindy Fan.
p. cm.
Includes bibliographical references and index.
1. Migration, Internal–China. 2. Rural–urban migration–China.
3. China–Social conditions. I. Title.
HB2114.A3F33 2007
307.2´40951–dc22
2007023208

ISBN10: 0-415-42852-1 (hbk)
ISBN10: 0-203-93737-6 (ebk)

ISBN13: 978-0-415-42852-1 (hbk)
ISBN13: 978-0-203-93737-2 (ebk)

For dad, Steven, and Philip

Contents

List of illustrations x
Acknowledgments xii

1 Migration, the state, and the household   1
2 Volume and spatial patterns of migration   19
3 The hukou (household registration) system   40
4 Types and processes of migration   54
5 Gender and household strategies   75
6 Migrants' experiences in the city   95
7 Impacts of migration on rural areas   117
8 Marriage and marriage migration   137
9 The Chinese migrant in the twenty-first century   162

Notes   173
References   180
Index   204

Illustrations

Figures

2.1	Provincial units and the three economic belts	28
2.2	Volumes of interprovincial migration within and among regions, 1990 and 2000 censuses	33
2.3	Interprovincial net migration, 1990 census	34
2.4	Interprovincial net migration, 2000 census	35
2.5	The 30 largest interprovincial migration flows, 1990 census	36
2.6	The 30 largest interprovincial migration flows, 2000 census	37
4.1	Dichotomies of migration reasons	56
4.2	Migration reasons, 1990 and 2000 censuses	62
4.3	Permanent migrants by migration reason, 1990 and 2000 censuses	63
4.4	Temporary migrants by migration reason, 1990 and 2000 censuses	63
5.1	Gender differentials in age-specific interprovincial migration rate, 1990 and 2000 censuses	78
5.2	Migration reasons by gender, 1990 census	81
5.3	Reasons for male interprovincial migration, 1990 and 2000 censuses	82
5.4	Reasons for female interprovincial migration, 1990 and 2000 censuses	83
7.1	Uses of remittances, 1995 Sichuan and Anhui Interview Records	119
8.1	Female net interprovincial marriage migration, 1990 census	151
8.2	Female net interprovincial marriage migration, 2000 census	152
8.3	Site of Gaozhou field study and origins of marriage migrants	153

Tables

2.1	Migration volumes, 1990 and 2000 censuses	21
2.2	Permanent migrants, temporary migrants and floating population, 1990 and 2000 censuses	23
2.3	Origin and destination types, 1990 and 2000 censuses	26
2.4	Interprovincial migration, GDP per capita and population, 1990 and 2000 censuses	29–30

2.5	Proportions of interprovincial migration within and among regions, 1990 and 2000 censuses	32
4.1	Migration reasons and definitions	55
4.2	Characteristics of migrants by migration reason, 1990 census	58
4.3	Summary comparison among migration reasons	59
4.4	Comparison among non-migrants, permanent migrants and temporary migrants, 1990 census	60
4.5	Logistic regression on permanent and temporary migrants, 1990 census (15+)	65
4.6	Comparison between permanent and temporary interprovincial migrants, 1990 and 2000 censuses	68
4.7	Migration considerations, 1998 Guangzhou survey	70
4.8	Migration considerations and information, 1995 Sichuan and Anhui Interview Records	72
5.1	Comparison between male and female migrants, 1990 and 2000 censuses	77
5.2	Division of labor in married migrant households, 1995 Sichuan and Anhui Interview Records	90
6.1	Job search, ownership sector, and job stability, 1998 Guangzhou survey	96
6.2	Definition of occupations	100
6.3	Occupational attainment of permanent and temporary migrants (15+), 1990 and 2000 censuses	102
8.1	Comparison between female "industry/business" and "marriage" migrants, 1990 and 2000 censuses	145–146
8.2	Husbands of interprovincial marriage and non-marriage migrants, 1990 census	150
8.3	Marriage intermediaries, 1999 Gaozhou field study	155
8.4	Age and education of wives and husbands, 1999 Gaozhou field study	157

Plates

5.1	Peasant women are expected to stay to take care of the children and farmland so that their husbands can pursue off-farm migrant work	91
6.1	Job advertisements in cities usually specify requirements of hukou, age, sex, and level of education	106
7.1	In some Chinese villages, most young men and single women have left for migrant work, leaving behind married women, their children, and the elderly	130
8.1	Villages in Gaozhou are characterized by a mix of older houses and new and bigger "lychee houses"	156

Acknowledgments

I would like to acknowledge the National Science Foundation (BCS-0455107; SES-0074261; SBR-9618500) and the Luce Foundation for funding the research on which this book is based. My sincere thanks to the review boards and anonymous reviewers for supporting the research, and to the NSF program directors, Thomas Baerwald, James Harrington, Jr., Richard Aspinall, Joseph Young and Gregory Chu for helping with various aspects of the grants. I am also grateful to the UCLA Academic Senate, Division of Social Sciences, Department of Geography and Asian American Studies Center, and in particular to Scott Waugh, David Rigby, Glen MacDonald, Denis Cosgrove, and Don Nakanishi for financial and administrative support for my research.

Early versions of some parts of the book were presented as keynote lectures and at a number of conferences and colloquia. I thank Larry Band for inviting me to deliver the J. Douglas Eyre Distinguished Lecture at the University of North Carolina, Chapel Hill, and Jianfa Shen and Jinn-yuh Hsu for inviting me to give keynotes at, respectively, the Fourth International Conference on Population Geographies held in Hong Kong and the Second Cross-Strait New Economic Geography Conference held in Taipei. I am thankful to the sponsors of the "Migration and Social Protection in China" Conference in Beijing, Workshops on "Urban China in Transition" in New Orleans and Santa Monica, the World Bank Workshop on China's 11th Five-Year Plan in Beijing, "The Future of U.S.-China Relations" Conference at the University of Southern California, the "Perspectives on Global Energy" Workshop in Houston, the "China at its Crossroads" Conference at Kutztown University, and the "Translocal China: Place-Identity and Mobile Subjectivity" Workshop in Haikou. My gratitude goes to Russell Smyth, Ingrid Nielsen, John Logan, Youqin Huang, Shahid Yusuf, Clayton Dube, Rachel Finch, Steven Schnell, Tim Oakes, and Louisa Schein for inviting me and to the conferences' panelists and discussants for their comments. Feedback from colloquia at Arizona State University, Miami University, UC Santa Barbara, University of Southern California, University of Washington, and University of Wisconsin-Madison has enriched my research for the book, and I would like to thank Robert Balling, Jr., Hong Jiang, Stan Toops, Laura Pulido, Kam Wing Chan, and Raymond Wong for their invitations.

Some parts of the book draw on papers previously published in *Annals of the*

Association of American Geographers, *Environment and Planning A*, *Eurasian Geography and Economics*, and *The Professional Geographer*. I thank Wiley-Blackwell, Pion Limited, London and V.H. Winston & Sons, Ltd. for permissions to adapt from illustrations in these publications. I am also thankful to the editors and reviewers of these papers for their helpful comments.

I am grateful to my collaborators whose contributions to a range of activities enriched my research for the book. I would like to recognize in particular Nansheng Bai for his longstanding commitments to and painstaking work on labor migrants in China, Kam Wing Chan for shrewd observations and seminal works on urbanization and migration, Ling Li for very effective coordination of fieldwork, Joanna Regulska for inspiring discussions, Allen Scott for his scholarship on and analyses of Asian economies, and Youqin Huang and Wenfei Winnie Wang for sharp insights and exemplary research. Several other former and current graduate students provided assistance to and gave helpful comments on parts of the book. Mingjie Sun deserves special thanks for prompt and very reliable data analyses. I also thank Jiantao Lu, Justin Zackey, Jenn Lee Smith, Chuncui Velma Fan, Ping Zhang, and Rui Yao for their able work. Chinese researchers and visiting scholars, in particular Huabo Wang, Wenxin Zhang, Daming Zhou, and Rong Zhang, gave invaluable research assistance, for which I am grateful. I am very thankful to the informants at the Gaozhou Statistical Bureau, Gaozhou Family Planning Commission and Genzhi Family Planning Commission, village residents in Gaozhou, respondents in Guangzhou and villagers in Sichuan and Anhui for their participation in field surveys and interviews that informed the book.

I am especially indebted to Laurence J.C. Ma and Mark Selden who read several chapters of the book closely and gave detailed and extremely helpful comments. Larry Ma and Clifton Pannell deserve special recognition for being my mentors for many years and for their deep commitments to and profound impacts on China geography and geographers. I also thank the boards and colleagues of the China Specialty Group of the Association of American Geographers for building a lively China geography community of which I am part. I am especially thankful to Jack Williams and C.P. Lo for their pioneering work and to George Lin, Greg Veeck, and Yehua Dennis Wei for their scholarship and friendship.

My sincere thanks also go to George Demko, David Goodman, Susan McEachern, Dorothy Solinger, and Fulong Wu for their advice on and support for the book project. The process of writing the book has been enriched by formal and informal discussions with many colleagues and friends, especially Shuming Bao, Lawrence Berg, Andy Bond, Fang Cai, Carolyn Cartier, Chen Chen, Delia Davin, Caroline Desbiens, Chengrong Duan, Mark Ellis, Kim England, Arianne Gaetano, Denise Hare, Chiao-min Hsieh, You-tien Hsing, Caroline Hoy, Tamara Jacka, John Paul Jones III, James Kung, Nina Lam, Vicky Lawson, Patrick Le Gales, Bingqin Li, Shi Li, Si-min Li, Wei Li, Zai Liang, Max Lu, Francis Lui, Zhongdong Ma, Margaret Maurer-Fuzio, Alison Mountz, John O'Loughlin, Marie Price, Ken Roberts, Rachel Silvey,

Christopher Smith, Mary Thomas, Andrew Waldor, Margaret Walton-Roberts, Feng Wang, Victor Winston, Weiping Wu, Yunyan Yang, Henry Yeung, Li Zhang, Zhenzhen Zheng, Shangyi Zhou, Yixing Zhou, and Yu Zhou.

My colleagues at UCLA are constant sources of inspiration and have supported my research in a variety of ways. John Agnew, Bill Clark, and Nick Entrikin gave incisive comments on an early version of the book proposal. Michael Shin gave timely and generous assistance on cartographic techniques, Judith Carney shared interesting materials on China with me, and Michael Curry's insights on writing were helpful. Don Treiman helped me navigate the complex review of human subjects protection, and discussions with him, Bill Mason, and Bill Lavely of the University of Washington, have helped tremendously the tasks of data management. Colleagues of the Center for Chinese Studies, especially Richard Baum, Benjamin Elman, Richard Gunde, Philip Huang, David Schaberg, James Tong, Bin Wong, and Yuxiang Yan, deserve much credit for promoting and supporting China Studies at UCLA. I also thank Katherine Bernhardt for her comments on an early paper. The research activities of my Asian American Studies colleagues, especially Lucie Cheng, King-Kok Cheung, Lane Hirabayashi, Jinqi Ling, Robert Nakamura, Don Nakanishi, Paul Ong, Tritia Toyota, and Min Zhou, have motivated me to view China from a more global perspective, for which I am thankful.

Many thanks also to the UCLA staff and administrators that helped with grant preparation and administration, especially Jason Corbett, Tina Schroeter, Kasi McMurray, Rebecca Goodine, Marc Mayerson, Christine Wang, Sharon Lam, Ken Castro-Oistad, Tana Wong, Joanne Yung, Charles Kim, Lois Erlanger, and Grace Lau. I would also like to recognize Michael Mitchell, David Deckelbaum, William Zhang, Aleks Gendeman, and Brian Won for providing technical support in computing. I am grateful to Chase Langford for his expertise and proficiency in producing maps for the book.

I thank Andrew Mould, Senior Editor at Routledge, for his enthusiasm about the book project. Many thanks also to Doug Jewsbury for his invaluable assistance in copy-editing and to Jennifer Page, Vicky Claringbull and Allie Waite for their efficiency in the book's production.

Finally, my deepest gratitude goes to members of my family for their love. I am grateful to my father, Sin Pong Fan, and my brother, Steven Fan, for their incredible support and for setting excellent examples for me. Philip Law, my husband, deserves special thanks for his patience and selflessness when I was preoccupied with the book, and his encouragement, humor, and kindness. To our children, Isaac and Naomi, I thank them for sharing their laughters and discoveries with me.

1 Migration, the state, and the household

Introduction

The rapid surge of migration has been one of the most profound changes in China since it embarked on economic reforms in the late 1970s. Official estimates of the floating population – people not living in places where they are registered – are in the range of 150 million, accounting for about 12 percent of China's population (National Bureau of Statistics (NBS) 2006). Some sources estimate that the number of migrant workers in Chinese cities is as high as 200 million, rivaling the total volume of international migrants worldwide (e.g. *China Daily* 2006).[1] The sheer size of migrant flows has profoundly affected China's development. Rural–urban migration, in particular, has been the main source of urban growth and is rapidly reshaping the economic, demographic, and social landscapes of the Chinese city and countryside. Parallel to, and as a result of, rural–urban migration, China's level of urbanization increased from 21 percent in 1982 to 43 percent in 2006 and is expected to exceed 50 percent by 2015 (Chan and Hu 2003; Duan 2003; Guangming ribao 2006; Lu and Wang 2006; Zhou and Ma 2003; 2005; see also Chapter 2).[2]

Research on migration in China, not surprisingly, has proliferated and has produced an impressive body of findings. Numerous studies have explored in depth the hukou (household registration) system and how it has given rise to two circuits of migration, one consisting of privileged migrants sponsored by the government and the other consisting of peasant migrants relying on their own resources (e.g. Chan and Zhang 1999; Chan *et al.* 1999; Cheng and Selden 1994; Fan 2002a; 2002b; Goldstein and Goldstein 1991; Gu 1992; Li 1995; Mallee 1996; Smart and Smart 2001; Wang 1997; Wang *et al.* 2002; Wu and Treiman 2004; Yang 1993). Related to this general theme is research on peasant migrants' inferior political, social, and labor market positions (e.g. Chan 1996; Fan 2001; 2002a; Jiao 2002; Smith 2000; Solinger 1995; 1999a; 1999b; Wang *et al.* 2002; Zhou 1992). Second, researchers have documented the patterns of migration, including spatial and regional patterns, migrants' characteristics, and different types of migration (e.g. Fan 1999; 2005a; Li 2004; Liang and White 1997; Liang and Ma 2004; Ma 1996; Scharping 1997). Third, some studies have focused on the determinants of migration and migration experiences, including economic opportunities, regional disparity, rural development, and network (e.g. Fan 1996; 2005b; Hare 1999a; Li 1997; Liang 1999; Rozelle *et al.* 1999; Shen 1999; West and Zhao 2000; Zhou 2002). Finally, the incorporation of migrants into the urban economy and society, including their occupations, housing, communities, fertility behavior and health, and the role of return migrants in the countryside, are increasingly popular research themes (e.g. Gaetano and Jacka 2004; Huang 2001; Jacka 2006; Li and Siu 2002; Logan 2002; Ma 2001; 2002; Ma and Xiang 1998; Murphy 2002; Pun 2005; Shen 2002; Smith 1996; Stinner *et al.* 1993; Taubmann 1997; Wu 2002; 2004; Yang 2006; Yang *et al.* 2007; Zhang 2001). Building on this fruitful body of work, this book seeks to highlight the roles of the state and the household for interpreting migration in China. I argue that the surge of migration and its changes must be understood in relation to how the state and the household have repositioned themselves during economic and social restructuring. In this chapter, I shall briefly establish the link between the state and the migrant labor regime and the role of gender in peasant household strategies. This is followed by an outline and critique of the hukou and permanent migrant paradigms. After a description of the data used in the analysis, I give a brief overview of individual chapters.

The Chinese state

Most mainstream migration theories do not emphasize the role of the state in internal migration. The neoclassical view emphasizes human capital and regional income differences as determinants of migration (Schultz 1961; Sjaastad 1962; Todaro 1969). The behavioral approach highlights psychological factors that constrain rational decision-making (Wolpert 1965). Network theory stresses the role of social networks and shared community origin in shaping the process and pattern of migration (Goodman 1981; Massey 1990). The labor market segmentation theory describes how migrants in cities are channeled into a secondary or informal sector (Harris and Todaro 1970; McGee 1982; Piore 1979). The household strategies perspective considers migration as part of the household's organization of resources (Mincer 1978; Katz and Stark 1986). Demographic explanations of migration focus on life-course triggers and household aging that create the need to move (Brown and Moore 1970; Clark *et al.* 1984). Compared to the above perspectives, those focusing on international migration, not surprisingly, are much more concerned with policy, legislation, and related institutional factors (Calavita 1992; Farer 1995; Massey and Espinosa 1997; Waldinger 1992; Wright and Ellis 1997; 2000).

In social sciences, the neoclassical and neoliberal positions that prioritize market over institutions are increasingly being challenged. Skocpol (1985), for example, calls for "bringing the state back in" to a central place in explanations of social change and politics. Amin (1999), Jessop (1999), and Peck (1994) are among those advocating an institutional approach in economic geography. Much of the debate about an institutional approach, including the role of the state, however, has taken place with reference to North American and Western

European contexts. The disconnect of the debate from transitional and formerly socialist economies is especially troubling because it is in these very economies that the state's role is most pronounced and has had the most profound changes. Russia and China share some similarities in the persistence of policies and institutions that aim at regulating migration and labor market processes. In Russia, Soviet-period institutions continue to control access to social services and benefits (Buckley 1995; Mitchneck and Plane 1995a; see also Chapter 3). Similarly, in China, peasant migrants' incorporation into the urban labor market is heavily monitored by the state. In both countries, internal migration is not free but accrues costs such as fines and refusal of social services, and labor markets have emerged only recently.

For almost three decades, China has undertaken a transition from the former socialist model into a "socialist market economy." Despite the adoption of market logic, the state continues to see itself as the ultimate planner that guides and regulates the economy. In that light, the "socialist" in socialist market economy refers less to political ideology but more to the domination of the Chinese Communist Party (CCP) and the very planning function of the state. Here, the state is broadly defined and includes not only the central government but also local governments and agencies and institutions authorized for planning purposes. Though the state does not necessarily have a unified and consistent approach all the time, its shift toward a developmentalist agenda is undeniable. This shift has greatly influenced how the state redefines its position, how it has engendered a migrant labor regime, its relations with peasants, and gender roles and relations – all central for understanding migration and in particular rural–urban migration in China.

The developmentalist state

Since the late 1970s, the Chinese state has taken on a developmentalist role. Losing faith in the Maoist model of central planning and collectivization and realizing that China's economic development was lagging considerably behind that of Newly Industrializing Economies (NIEs) (South Korea, Taiwan, Hong Kong, and Singapore) and Western capitalist economies, Deng Xiaoping and his associates were determined to adopt measures that boost economic growth. Deng set the tone for economic reforms by the famous quote "It doesn't matter if the cat is white or black so long as it can catch mice," which epitomizes the shift in focus from political ideology to pragmatism, expertise, and performance. Policies that would have been unacceptable during Maoist times, such as marketization, decollectivization and fiscal decentralization, have become appealing because of their promise to bring about economic growth.

The mandate of rapid economic growth has also legitimized a strategy of export-oriented industrialization, pursued through an open-door policy, export-processing and special economic zones, and incentives for foreign investors. Chinese scholars accept this strategy as applications of the "grand international cycle" theory, which in essence describes capital's global search for new, cheap sites of investment (Cheng 1994). Past policies that emphasized agriculture have virtually been abandoned and replaced by a new focus on modernization via industries and services (Lo 1994; Yang and Guo 1996). Clearly, the new development strategy was inspired by the success of NIEs, but it was also made possible by the abundance of rural labor. In short, adoption of export-oriented, labor-intensive industrialization demands a labor regime different from that of the Maoist period, one that depends heavily on migrant workers.

Migrant labor regime

The state's approach toward labor, during the Maoist period, was one based on centralized allocation. Through "unified state assignment," state agencies allocated jobs to school graduates and transferred workers from one job to another. This approach entailed low job mobility and controlled labor migration. In this connection, the People's Republic of China's (PRC) version of the hukou system was implemented in the late 1950s (see Chapter 3). Under this system, urbanites were entitled to work and access to subsidized food, housing, education, and other social services. Open markets for food, housing and jobs were virtually non-existent, and almost all necessities in urban areas were controlled by the state. Without urban hukou and accompanying benefits, it was next to impossible for peasants to survive in cities. Thus, peasants were bound to the countryside and rural–urban migration was minimal.

While the hukou system was an instrument of migration control (e.g. Yu 2002: 15–21), it has also been seen as a tool for implementing the Maoist state's development philosophy. Specifically, the state blocked flows of resources, including labor, from rural to urban areas in order to extract value from agriculture for subsidizing industry – especially heavy industry – as it pursued a Soviet-style development model. By binding a large labor force to the countryside, this strategy ensured a supply of low-priced agricultural goods to achieve "industrialization on the cheap" (Chan and Zhang 1999; Cheng and Selden 1994; Tang *et al.* 1993). The premise of this political-economic argument is that the state erected barriers between the city and the countryside in order to advance a specific model of development (see also Chapter 3).

A similar logic is useful for understanding the hukou system and its changes during the post-Mao period. Far from being dismantled, urban–rural barriers have been used by the developmentalist state to achieve new economic goals. By maintaining an institutional and social order in which peasants are inferior to urbanites, and by permitting peasants to work in urban areas as "temporary" migrants – migrants that are denied urban hukou and entitlements – the state has created a migrant labor regime that enables labor-intensive industrialization and urban development (see Chapter 3). In this way, the state makes available a large supply of rural labor to advance its developmentalist strategy at low cost and at the same time ensures that most peasant migrants will eventually return to the countryside without burdening the state.

These peasant migrants are attractive to global investors. Peasants' institutional and social inferiority, as well as severe labor surplus in the countryside, are the key to explaining their pursuit of urban jobs despite the low pay. Without social insurance and labor rights infrastructure, the migrant labor regime is a safe haven for urban and industrial employers that thrive on cost-minimization and exploitation (see Chapter 6).

On the surface, the migrant labor regime in China is not too different from the labor market in many developing countries, where a cheap and unskilled labor force fosters labor-intensive industrialization. The labor market segmentation theory, in particular, explains the channeling of migrants into the secondary, informal sector in cities. This theory, however, assumes homogeneity among migrants and does not highlight the role of the state (Breman 1976). What makes China stand out is the central role of the state in channeling and constraining peasant migrants to specific sectors and jobs – construction, garment factories, domestic work, and other jobs shunned by urbanites – through control instruments in connection with the hukou system. Temporary migrants in Chinese cities are not spillovers from the primary sector; rather, they are blocked by state institutions from entering the primary sector. The migrant labor regime is, in essence, the product of a system that defines opportunities by hukou status and locality and that fosters a deep divide between rural and urban Chinese. Despite hukou reforms that have taken place since the late 1980s (see Chapter 3), the vast majority of peasant migrants continue to be in inferior institutional, economic and social positions compared to urban residents.

The migrant labor regime is also gendered. Similar to many other parts of the world, the availability of young, single women workers in China has attracted investment from multinational corporations (e.g. Chant and Radcliffe 1992; Cheng and Hsiung 1992). Factories target young, single migrant women because they are construed as having good attention to detail, able to handle delicate work, and easy to control (e.g. Lee 1995), while male migrants are channeled to heavy work such as construction. Thus, migrant work is highly segregated by gender (Fan 2003; Yang and Guo 1996). In the migrant labor regime, new forms of labor disciplining that are conducive to the penetration of global capital are also gendered (Ong 1999: 38–40).

To be sure, the migrant labor regime is not the only labor regime in post-Mao China. Remnants of the socialist period based on state labor-allocation, and greater job mobility of skilled and professional workers, also characterize China's labor market. The migrant labor regime is, however, a profound testimony to the state's role in using socialist control instruments such as hukou to foster its new developmentalist goals.

The peasant household

The state and the peasants

"To understand the real forces driving migration in China, it is necessary to look to the countryside" (Veeck *et al.* 2007: 121). In contrast to the Maoist period, when peasant households were part of communal production, the

6 Migration, the state, and the household

Household Responsibility System (HRS) that was formalized in the early 1980s and subsequently adopted throughout the countryside returned decision-making to the household. Peasant households contract from the village authority farmland that is allocated primarily according to household size. This new system is credited for boosting agricultural production (Lin 1992). It has directly affected peasants' lives in two ways. First, improvement in agricultural productivity further exacerbated the problem of surplus labor that had been hidden in the form of underemployment during the collectivized period. Population growth, in addition, increased the magnitude of rural labor surplus and exerted pressure for peasants to find off-farm work (Banister and Taylor 1989). This five-person household (husband, wife, son, daughter, and husband's mother) in Anhui, for example, was allocated 7.5 acres of farmland in 1980 when the HRS was adopted in their village, but saw their contract land shrinking over time: "in 1994, the village conducted a major land reallocation. Because of population growth, per capita farmland [in our village] was reduced from 1.5 acres to 1.3 acres. Our family's allocated land declined [from 7.5 acres] to 6.5 acres" (NNJYZ 1995: 349–350). Decline in the land-to-person ratio is felt across the Chinese countryside, where significant proportions of labor are idle. Estimates of the size of agricultural labor surplus are generally in the range of 150 million to 200 million and are expected to increase (Cai 2001; Cao 2001; Jian and Zhang 2005; Meng 2000; Qiu 2001; Wang 2001; Zhang and Lin 2000). The sentiment that young people are not needed in agriculture is widespread and is one frequently reported in surveys (see Chapter 5).

Second, abolition of communes signifies the state's increased disengagement from peasants and from agricultural work. Chinese peasants are poor to begin with. Despite opportunities for improving agricultural production made possible by economic reforms, the burden of financing local economic development is increasingly put on peasant households. This is due largely to the developmentalist state's emphasis on urban and industrial development and its detachment from rural production (Wu and Ma 2005). High taxes and fees charged by local governments, in conjunction with the rent-seeking activities of cadres, make rural life extremely difficult. Wu's (2000) study of a village in Zhengzhou, Henan, for example, underscores the stratification between cadres and entrepreneurs on one hand and deprived villagers on the other. Even though the state has gradually removed the caps over prices of agricultural goods, agriculture remains a poor source of livelihood. Studies have shown that there is a strong desire to leave agriculture (Croll and Huang 1997). To many peasants, income from agriculture and other sources in the countryside is barely enough, if not outright insufficient, to make ends meet. Compared to the pre-reform period, and unlike urbanites, peasants have little access to state support and must rely on their labor, contract land, and household strategies for survival and economic mobility, as summarized by this Sichuan woman: "Since the beginning of contract farming, we are all on our own [compared to collective farming]" (NNJYZ 1995: 85–86). Under these circumstances – prospect of persistent poverty, large labor surplus, and removal of communal protection and state support – migrant work has become an attractive strategy for augmenting peasant households' income and improving their well-being (Croll and Huang 1997; Tan 1996a). The term *dagong* – literally "being employed" and referring specifically to rural people seeking work elsewhere, mostly in industrial and service sectors – describes this phenomenon. Young female migrant workers are often referred to as *dagongmei* and young male migrant workers *dagongzai*.

Despite the economic gain from migrant work, protecting and taking care of the farmland continues to be a high priority for most peasants. This reflects the importance of rural livelihood to peasant migrants (see Chapter 5), as well as the absence of a market for farmland. Specifically, Chinese peasants are not free to buy or sell farmland, transfer of their leases to others is possible but not encouraged, and conversion of farmland for other purposes is strictly controlled by the state (Cai 2000; Lin and Ho 2005; Yusuf and Nabeshima 2006: 57–58). To some extent, therefore, they are similar to Mexico–US immigrants who have access to farmland in rural Mexico that cannot be sold and that has become not only an economic asset but also a base for all household activities (Roberts 2007). In rural China, likewise, peasant migrants commonly leave behind household members to farm. Other, albeit less popular, arrangements include asking siblings or relatives to take care of the farmland or leasing the farmland to others until the migrant returns. This 35-year-old Sichuan man, who has worked as a loader in Shenzhen for over ten years, explains:

> There are two advantages to leasing the farmland to others. First, they will protect the farmland's boundaries [so that it will not be taken over by other villagers]. Second, it [having someone take care of the farmland] will prevent the growth of wild plants.
>
> 2005 Sichuan and Anhui Interview Records

Household strategies

The household as a unit of analysis has increasingly gained popularity in social sciences. Wallace (2002) argues that household strategies are likely to become more important when a society is subject to rapid social change that leaves households in a situation of risk and uncertainty, when more women enter into the labor force, and when large parts of the economy are informal. All three conditions describe current situations in China.

Studies on migration have shown that household considerations and strategies are central to explaining migration decisions, patterns and outcomes. The conventional household strategies approach emphasizes the economic or utility gains households can make via migration (Boehm *et al.* 1991). An influential view is that net family gain, rather than net personal gain, motivates migration. Thus, migration is often interpreted as a strategy to increase income and diversify income sources for the entire household (Adams and Page 2003;

De Jong 2000; Ortiz 1996). In some developing countries and poor areas, migration and remittances from migrant work are essential for the subsistence of rural households and households in poverty (Goldscheider 1987; Itzigsohn 1995; Radcliffe 1991).

An approach that focuses on economic calculations, however, tends to pay little attention to the social relations and hierarchy that underlie household decision-making. Likewise, most studies on migration in China highlight the economic reasoning of household strategies rather than the power dynamics among household members (Hare 1999b). Increasingly, however, researchers have begun to highlight the non-economic factors of migration decisions (Clark and Huang 2006; Hugo 2005; Zhao 1999). Odland and Ellis (1988), for example, find that potential migrants may forgo the economic benefits of migration in order to keep the household intact.

Recent advancements to the household strategy approach, especially by feminist scholars, draw attention to intrahousehold power relations, including gender hierarchy and migrants' agency (Chant 1996; Chant and Radcliffe 1992; Eder 2006; Fincher 2007; Jarvis 1999; Lawson 1998; Lingam 2005; Momsen 1992; Silvey and Lawson 1999; Willis and Yeoh 2000). Marriage and gender roles within marriage are, in particular, found to be key to explaining differences in migration process and decision-making between men and women (Cerrutti and Massey 2001; Radcliffe 1991). A feminist approach to household strategies, likewise, can shed important light on migration decision-making in the Chinese peasant household. This book's approach is also informed by feminist methodologies that emphasize understanding people's lives, which enable more equalized power relations between the researcher and the individuals s/he studies, facilitate a bottom-up approach in intellectual reasoning that challenges the assumption that global and national forces are preemptive, and add richness and texture to the otherwise abstract theorization of the questions at hand (Nagar *et al.* 2002; see "Data").

In China, the age-old concept of the family (*jia*) is key to understanding how members of a household make mobility decisions. Members of a family are related to each other by blood or marriage and their budgets, properties, and interests are interconnected, even though they may live in different households and even separate places (Woon 1994). Thus, household strategies often include family members not residing under the same roof. Increased migration does not appear to have undermined the concept of extended families (Goldstein *et al.* 1997). Thus, a migrant's decision-making almost always involves considerations for and input of other family/household members (Rowland 1994). For example, a peasant migrant's income benefits the entire household, while household members that stay behind facilitate migrants' eventual return. Split households are extremely popular in rural China, most frequently involving the husband doing migrant work and the wife staying in the countryside (Tan 1996a; see Chapter 5). Inasmuch as migration involves the collaboration of, and division of labor among, different household members, the social and hierarchical power relations within the household are crucial for understanding decision-making and outcomes of migration (Lawson 1998; Radcliffe 1991). Gender roles and relations, to which I shall turn, are central to these relations.

Gender

The Chinese traditional view of gender is one rooted in Confucianism, which prescribes individuals' roles based on their positions relative to others. Accordingly, the Chinese woman is defined in relation to, and subordinate to, other males in the family – she is supposed to submit to her father before marriage (*zaijia congfu*), to her husband within marriage (*chujia congfu*), and to her son(s) during her old age (*laolai congzi*). This patriarchal ideology underlies two traditions – the inside–outside dichotomy and patrilocal exogamy – which persistently undermine women's status in China. The inside–outside dichotomy defines the woman's place to be inside the family and the man's sphere to be outside (*nan zhu wai nu zhu nei*) (Mann 2000). "Men till, women weave" has long been the model for gender division of labor in the countryside (Entwisle and Henderson 2000: 298; Hershatter 2000). Ironically, Chinese women have indeed made significant contribution to farm work, but the "men till, women weave" rhetoric has hidden their contribution to agriculture (Hershatter 2000: 83). Rural–urban migration in recent decades has further increased women's responsibility in agriculture, as many are left behind to farm while their husbands pursue migrant and off-farm work (Stockman 1994). Thus, as argued by Jacka (1997), the boundary of the inside, women's sphere has expanded beyond the physical interior of the family compound to including agriculture (see Chapter 5).

Second, Chinese marriages, especially rural marriages, are governed by patrilocal exogamy, whereby the wife moves out from the natal family and joins the husband's family. This tradition is key to explaining women's low status because the eventual loss of daughters discourages the natal family from investing in their education relative to their male siblings (Li 1994a; Lu 1997). This view, even today, determines how young women in the countryside are treated, as illustrated by this Sichuan woman's comment on migrants' remittances: "Daughters married out are like water spilled out. They don't usually send their remittances to the natal family. Even if they do, it's largely symbolic. Sons are different. They are expected to shoulder the responsibility for the family" (2005 Sichuan and Anhui Interview Records). Patrilocal exogamy also explains early marriage, because the husband's family is eager to recruit the labor of the daughter-in-law and because the natal family is interested in the daughter's finding a mate before she gets "too old" (see Chapters 5 and 8).

The extent to which the Maoist state had improved gender equality is a subject of heated debate. Mao's famous statement that "women carry half of the heavens on their shoulders" emphasizes women's contribution to economic production. Indeed, women's labor force participation rate in China has significantly increased and is similar to those in Western industrialized economies (Riley 1996). In addition, through legislation such as the 1950 Marriage Law, women's rights were incorporated into the legal codes, and feudal practices such as arranged marriages and footbinding were outlawed (Croll 1984). Critics, however, argue that the Maoist version of feminism focuses on the sameness between men and women but does not address, acknowledge or alleviate women's responsibility for housework and motherhood and does not challenge patriarchy (Harrell 2000; Hershatter 2000; Johnson 1983; Smith 2000: 290). Studies have shown that the Maoist state did not succeed in equalizing gender opportunities in education, labor market, and Chinese Communist Party (CCP) membership (Bian *et al.* 2000).

Evaluations of women's status during the post-Mao period are, likewise, mixed. The 1981 Marriage Law and its revisions in 2001 expanded further the rights of women in relation to marriage, divorce, and property. In 1995, the United Nations Fourth World Conference on Women was held in Beijing, apparently giving a boost to collective feminist awareness. Concerns over mistreatment of women and other gender issues have gained visibility in public discourses. Increasingly, Chinese intellectuals are publishing papers and books on gender roles and relations. The term "gender," which does not have a direct equivalent in the Chinese language and is now commonly translated as "social sex" (*shehui xingbie*) – emphasizing the social contexts in which gender identity is constructed – is increasingly recognized and used.

However, by retreating from an explicit gender policy and pursuing a developmentalist agenda, the post-Mao state is seen to have promoted patriarchy and undermined women's interests (Park 1992). The one-child policy, in particular, legitimizes the state's surveillance of women's bodies, invades their privacy, and penalizes fertility (Smith 2000: 315). Distorted sex ratios and systematic evidence of "missing girls" reveal the prevalence of sex-selective abortion, girl infanticide, neglect and underreporting of girls, especially in the countryside (e.g. Banister 2004; Li and Lavely 1995; Riley 1996; Smith 2000: 313–316; Xu and Ye 1992; Whyte 2000). Decollectivization in agricultural production is, likewise, gendered, as the return to household-level decision-making reinforces the husband's position and women's subordination (Smith 2000: 309–311). Numerous studies have reported that women's position in the labor market has deteriorated, that they are discriminated against in hiring and career development, and that they are segregated into "female" sectors in both urban and rural areas (Bian *et al.* 2000; Goldstein *et al.* 2000; Honig and Hershatter 1988; Jacka 1997: 193; Lin 2000; Maurer-Fazio *et al.* 1999; Wang 2000; Yang 2000b). The notion that women have lost ground during the reform period is, however, not unchallenged. Whyte (2000) questions the empirical soundness of comparisons between the Maoist and post-Mao periods. Wang (2000) is skeptical about the notion that farming is increasingly engaged by women. Michelson and Parish (2000) show that women in more developed locations are able to "catch up" via non-farm employment.

Gender roles and unequal gender relations governed by patriarchal ideology are fundamental to the formulation of household strategies in relation to migration. In the countryside, women have low educational attainment, marry young, and are designated for care-giving, house chores, and farming. Their roles are essential to the success of a split-household strategy and the husband's eventual return from migrant work (see Chapter 5).

The hukou and permanent migrant paradigms

In the literature on migration in China, two parallel paradigms have emerged that have deeply influenced how researchers understand and interpret rural–urban migration in China. The first focuses on the effects of the hukou system; the second emphasizes the desire of peasant migrants to enter and stay in cities and of temporary migrants to become permanent settlers. Both have influenced and guided many insightful studies, but they do not fully explain population movements in China. Below, I discuss the two paradigms and some of their inadequacies.

Hukou paradigm

It is now widely accepted that the hukou system has played an important role in producing and reinforcing divides between the city and the countryside and between rural and urban Chinese. A dominant theme in the literature is that the hukou system has engendered a dualist structure such that the vast majority of peasant migrants have inferior statuses, entitlements, treatments, and opportunities compared to urbanites. Here, I shall not go into the details of this line of argument, but Chapters 3, 4 and 6 of the book focus, respectively, on the hukou system, its impacts on migration, and its relationship with the labor market.

No doubt, the hukou system is a primary reason for the disadvantaged institutional and social positions of peasant migrants. In this book, I acknowledge, document and analyze the impacts of the hukou system on population movements, the labor market, and the experience of peasant migrants. At the same time, I argue that an approach that emphasizes the hukou system alone risks privileging top-down, structural and institutional explanations over bottom-up, household and individual-level perspectives. I show that the strategies developed by peasant households and the ways in which intrahousehold power and social relations underpin these strategies are central to understanding migration. Studies using the hukou paradigm perspective assume that removal of the hukou system or awarding urban hukou to peasant migrants will alleviate peasant migrants' hardship and reduce the inequality between rural and urban Chinese (see Chapter 7). But, to many peasants, obtaining urban hukou may not be as appealing as household division of labor that permits one or more members to circulate between and straddle rural and urban economies. The strategy of engaging in activities in both rural and urban areas, meanwhile, increases household income and diversifies risk. According to the New Economics of Migration (NEM) theory, remittance is part of an implicit contract between the migrant and the household that is grounded on attachment to the origin community and a plan for eventual return (Lucas and Stark 1985; Stark and Lucas 1988). In this book, while fully engaging in analyses of the role of the hukou system, I am highlighting also the importance of the household and its strategies for interpreting migration in China.

Permanent migrant paradigm

In the general literature on migration, the conventional approach toward understanding temporary migrants is that they desire to stay. The experience of guest-workers in post-war Europe who eventually developed permanent communities, for example, supports the notion that temporary migration is a prelude to permanent settlement, as aptly summarized by the oft-repeated phrase "there is nothing so permanent as a temporary migrant." This permanent settlement paradigm, however, is increasingly being challenged, especially in the context of international migration and transnational communities (Saxenian 2005). Hugo (2003a; 2006) argues that while non-permanent and circular migration has increased rapidly, migrant workers do not always desire to settle in destination countries. Modern forms of transport and communication have reduced the friction of distance and allowed migrants to maintain closer and more intimate linkages with their home countries and communities than before. In addition, migrants can obtain the best of both worlds by earning in high-income destinations and spending in low-cost origins. By keeping their family at the origin, migrants can maintain valued traditions and family ties and make frequent visits. While in the past immigrants were expected to apply for citizenships and commit themselves to the host country, now dual citizenships are common and are recognized by more than half of the world's nations (Clark 2007). Concepts of international circulation of labor and international labor markets are, therefore, increasingly relevant.

Within countries, likewise, circular migration is on the rise. Temporary migration has always been common in Africa and Asia (Nelson 1976). In addition, rural–urban circular migration is the fastest-growing type of temporary migration in countries that are experiencing rapid urbanization and industrialization, including Vietnam, Cambodia and China (Deshingkar 2005). Persistent circular or seasonal migration within countries or between neighboring countries is emerging as the migration pattern of the poor.

The prevailing assumption about rural labor migrants in China is that they desire to stay in the city and bring their families there but are unable to do so because of the hukou system. Some studies have, however, found that the desire for peasant migrants to settle in cities is not as strong as expected and that the majority wish to return (Cai 2000: 145; Hare 1999a; Solinger 1995). Other studies show that many migrants choose not to obtain urban hukou even if given the opportunity to do so. Despite the aggressive hukou reform in Shijiazhuang since 2001, for example, only a small fraction of migrants took advantage of the reform and changed their hukou to the city (Wang 2003). Zhu's (2003; 2007) surveys in Fujian conducted between 2000 and 2002 found that only small proportions of the

Migration, the state, and the household

floating population would move the whole family to the city even if their hukou could be freely transferred. While there is little systematic evidence to explain why peasant migrants would refuse urban hukou, the answer probably varies considerably depending on the specific city's labor market, the home village's resources and locations, and the household's economic and social situation. In this book, I wish to show that peasant migrants have choices and that these choices – between marginal existence in the city and returning to the countryside, for example – are made possible by peasants' mobility and circularity. Put in a different way, I argue that temporary migration enables peasant households to advance economically by obtaining the best of both worlds and that this strategy does not necessitate permanent settlement in the city.

Data

Analyses in this book are based on national census data, survey and field data, and narratives. A multiple-source approach facilitates the triangulation of observations, especially since the data on migration in China are scattered, unstandardized, and varied in scale. Unlike single-source studies that address only the origin or destination of migrants, I examine migrants' experiences at both ends and in multiple locations in China. Furthermore, these multiple sources permit both quantitative and qualitative analyses. Macro-level data such as the census facilitate the quantitative documentation and description of general patterns. Survey and field data, on the other hand, may not be representative but are valuable for explaining and interpreting the patterns observed from macro-level data. Qualitative materials such as narratives, in particular, add depth to the examination of processes and shed light on the meanings of structural forces at the household and individual levels. Personal narratives are powerful means for identifying migrants' agency and strategies, and they enable a bottom-up research approach that can bring to the foreground the voices and experience of marginalized individuals in society (Jacka 2006: 10; Nagar *et al.* 2002). In short, my strategy is to employ macro-level quantitative data to describe overall patterns and changes and to use field and qualitative materials to exemplify some of the processes underlying these patterns and to underscore agency, negotiation, and conflicts. In this way, I combine the strengths of different forms of data. In the book, citation of narratives and related qualitative materials is done in such a way that ensures respondents' anonymity – their names are withheld or only pseudonyms are used, and the exact names and locations of respondents' home villages are not revealed.

1990 and 2000 censuses

Both the 1990 and 2000 censuses include systematic information about migration and are by far the most comprehensive national-level sources of data on population movements in China (see Chapter 2). Unless otherwise specified, analyses in this book are based on data from the printed publications by the

14 *Migration, the state, and the household*

National Bureau of Statistics (NBS) (formerly State Statistical Bureau, SSB) and from two micro samples (NBS 2002; SSB 1992; 1993).

The first micro sample is the 1990 census one percent sample. It is a clustered sample including every individual in all households of the sampled village-level units (village, town, or street) and containing all the information from the census form.[3] All figures presented in this book that are based on this sample have been multiplied by 100.

The second micro sample is the 2000 census 0.1 percent interprovincial migrant sample, based on the long form. The 2000 census is China's most recent census and is the first census that uses both a short form and a long form. The short form includes basic demographic information, while the long form contains most of the questions relevant to migration. The long form was sent to 10 percent of the households, yielding a sample accounting for 9.5 percent of the total population (NBS 2002: 1; Wang and Ye 2004). Thus, data presented in this book that are based on the 0.1 percent interprovincial migrant sample have been multiplied by 1,000/0.95.[4] Because this sample includes only interprovincial migrants, comparison between the 1990 and 2000 censuses in this book focuses primarily on interprovincial migration and not intraprovincial migration.

1998 Guangzhou survey

In addition to census data, I use several survey and field sources to examine in particular processes and decision-making related to migration. The first source is a questionnaire survey in Guangzhou that I conducted in June and July of 1998,[5] which informs the discussion of migration considerations and labor market experiences respectively in Chapter 4 and Chapter 6. Guangzhou is a major magnet for migrants of all kinds, from other parts of Guangdong and from other provinces; it has a large, diverse and changing economy and a rapidly expanding urban labor market. For these reasons, Guangzhou is among the best field sites for studying the relationship between migration and the urban labor market. The survey includes three types of respondents – 305 non-migrants, 300 permanent migrants, and 911 temporary migrants. A large number of temporary migrants were included because they are the newest and most dynamic migrants in cities and because censuses have likely underestimated the volume of temporary migrants (see Chapter 2). The sample was arrived at using stratified quota sampling, with stratification both across occupational categories and districts in Guangzhou. The sampling process and framework are detailed elsewhere (Fan 2002a) and are therefore not repeated here.

1999 Gaozhou field study

In 1999, I conducted fieldwork in Gaozhou, a rural area in the western periphery of Guangdong.[6] The main objective of the field study is to examine the relationships between marriage, marriage migration, and labor migration. Both a receiving area of marriage migrants and a sending area of labor migrants, western

Guangdong as a field site is revealing of the ways in which social, gender and power relations within households shape decision-making and processes of migration. The field study consists of a questionnaire survey and in-depth interviews. In Chapter 8, which focuses on marriage and marriage migration, I describe details of the site and the field study.

1995 Sichuan and Anhui surveys (Household Survey and Interview Records)

Sichuan in western China and Anhui in central China are two major sources of rural–urban labor migrants, especially to eastern and southern parts of the country (see Table 2.4 and Figures 2.3 to 2.6). Both provinces are among the poorest in China and have large volumes of rural surplus labor. In this book, I use information from three sets of surveys conducted in the two provinces.

The first set of surveys is the 1995 Sichuan and Anhui surveys, which were conducted by the Research Center for the Rural Economy (RCRE) of the Ministry of Agriculture (Du 2000; Du and Bai 1997: 6; Fan 2004b; 2004c). The surveys include two parts. The first part – 1995 Sichuan and Anhui Household Survey – involves a questionnaire survey of 2,820 households that were drawn from the survey database of NBS's Rural Social Economic Survey Team (RSEST) (*nongcun shehui jingji diaocha zhongdui*).

The second part – 1995 Sichuan and Anhui Interview Records – involves in-depth interviews with 300 households. The households were drawn from 12 villages – three villages each from two counties in Sichuan and two counties in Anhui. The counties and villages were selected based on the following criteria: that in terms of economic development they are representative of the respective provinces; that they have been sending out labor migrants for quite some time; and that labor out-migrants account for at least, respectively, 20 percent of the county's labor force and 30 percent of the village's labor force (Du and Bai 1997: 5). In each of the villages, 15 migrant households (where one or more members has had migrant work experience) and ten non-migrant households were randomly selected. The interviews were conducted during the Spring Festival in early 1995 – a time of year when many migrants return to the home village. The Interview Records includes a total of 191 migrants, of whom 83.8 percent are men. Interviewees' responses are in the form of narratives and are concerned with migration and labor market processes, farming, evaluation of migrant work experience, the household and the family, and views on a variety of issues (NNJYZ 1995). Each of the accounts was transcribed verbatim. Most narratives are more than 3,000 (Chinese) words long.

1999 Sichuan and Anhui surveys (Household Survey and Interview Records)

The second set of surveys on Sichuan and Anhui were conducted in 1999 also by RCRE (Bai and Song 2002: 10, 14; Wang and Fan 2006). Unlike the 1995 surveys,

16 *Migration, the state, and the household*

which focused on labor out-migration, the 1999 surveys' main concern was return migration. Like the 1995 surveys, the 1999 surveys consisted of two parts. The 1999 Sichuan and Anhui Household Survey was a questionnaire survey of 5,484 households, also drawn from RSEST's database. The second part – the 1999 Sichuan and Anhui Interview Records – involved in-depth interviews with 305 households and 39 return entrepreneurs (Bai and Song 2002: 11). The households were drawn from 12 villages – three villages each from two counties in Sichuan and two counties in Anhui. The counties were selected from those that had high rates and relatively long histories of labor out-migration, were representative of rural areas and were geographically spread out in the respective provinces (Bai and Song 2002: 10). In addition to these criteria, villages were selected to represent areas of varied levels of economic development within the respective counties. In each of the villages, 15 return migrant households (households with one or more return migrants), five migrant households and five non-migrant households were randomly selected. The interviews took place in May and June of 1999. Again, the responses were transcribed verbatim and they are mostly more than 3,000 (Chinese) words long. While the narratives deal primarily with return migration, they also address various aspects of economic and social production in the countryside (NNJYZ 1999).

In this book, I draw more from the 1995 and 1999 Interview Records than the respective Household Surveys, because the former provide first-hand, qualitative accounts that are uniquely suitable for drawing attention to the richness and texture of migrants' experiences. Selected narratives from the Interview Records that reveal salient aspects of individual and household-level processes and migration and labor market experience are used throughout the book.

2005 Sichuan and Anhui Interview Records

The last set of surveys on Sichuan and Anhui consists of interviews with the same 300 households originally interviewed in 1995.[7] This is part of a collaborative project that involves the author and the Renmin University of China. The purpose of re-interviewing the same households ten years after is to document and explain changes that have taken place, including, for example, whether peasant migrants are more attracted to and more able to settle permanently in cities than before, what impacts hukou reforms have had on migration decisions, and how new and younger migrants compare to their predecessors. Like the previous two Interview Records, each of the accounts is transcribed verbatim. At the time of this book's writing, the project is ongoing and only some of the transcribed materials have been analyzed. Therefore, in the book, I examine only narratives from interviews conducted in two villages in Sichuan and I use them primarily in Chapter 9 to address the most recent changes.

Overview of chapters

Chapter 2 gives an overview of the changes in the volume, rates, and spatial patterns of migration in China over the past two decades or so. Using mainly data from the 1990 and 2000 censuses, I describe flow and stock measures of migration, and I show that migration volume, especially that of temporary migration, has risen sharply since the mid-1980s. The shares of interprovincial and interregional migration have also increased, depicting decline in the friction of distance. Over time, as migration flows become more concentrated, more developed provinces in the eastern coast are gaining more migrants, whereas poorer provinces, especially those in the central region, are losing more migrants.

Chapter 3 focuses on the hukou system. It begins by describing hukou type and hukou location and their functions. After outlining prevailing explanations for the implementation of the hukou system, I discuss how the system operated in conjunction with central-planning mechanisms that allocated food, employment and welfare, resulting in tight control over rural–urban migration during the Maoist period. Then, I review the main criticisms toward the hukou system and major components of hukou reforms since 1984. Despite these reforms, hukou continues to constrain Chinese peasants' access to opportunities and resources.

In Chapter 4, I argue that both the conventional economic versus social dichotomy and the state-sponsored versus self-initiated, market-driven dichotomy are useful for interpreting population movements in China. Over time, the roles of economic motives and market-driven processes have gained importance. I identify the relationships between different migration types on one hand and the two-track migration system on the other. The analysis shows that hukou is not only a reward to individuals' human capital and competitiveness but also a gatekeeping tool that allows only those sponsored by the state to achieve permanent migration. Relying heavily on social networks for information, temporary migrants' overriding objective is income gain.

Analyses in Chapter 5 challenge conventional notions about gender differentials in migration. The number of female migrants in China is large and increasing. They travel long distances and their moves are increasingly driven by economic motives. However, deep-rooted patriarchy continues to govern gender roles and relations, such that peasant women's economic mobility via migrant work is short-lived. Married women's staying in the village reinforces the age-old inside–outside ideology but also enables the split-household strategy, thus facilitating husbands' circular migration, economic production at both the rural origin and the place of migrant work, and eventual return. Gender and household division of labor makes it possible for migrants to straddle the city and the countryside and for some to obtain the best of both worlds.

Chapter 6 examines the labor market and social positions of peasant migrants in cities. I show that both employment practices and social networks channel peasants to jobs marked by blatant exploitation. Segmentation of the urban labor market results not only from human capital and skills differentials but also from institutional and social exclusion. Peasant migrants are seen and treated as outsiders to the urban economy and society. In light of this, they tend to pursue monetary remuneration rather than long-term residence and they tend to solicit support from native-place networks and not from the host area.

Focusing on the countryside, Chapter 7 shows that labor out-migration has clearly had positive impacts on rural income and productivity. Remittances, in particular, enable peasants to build houses and support key aspects of rural living, which in turn facilitate migrants' return in the future. Migration's effect on agriculture and the impacts of return migration are, however, less conclusive. The social impacts of migration are mixed. Evidence points to both entrapment of married women in the countryside and adoption of urban views that are potentially critical of traditional gender and social hierarchy.

In Chapter 8, I show that marriage is employed by peasant women – especially those having the least competitiveness in the labor market – to move to more prosperous parts of the countryside. I discuss in detail the ideology and practice of marriage in China, and I review the patterns of marriage migration and the characteristics of marriage migrants. The findings highlight the roles of attribute matching and trade-off between female marriage migrants and the prospective husbands, social and kinship networks, peasant women's agency and strategies, and location. Marriages resulting from migrant work, however, appear to be based more on affection and less on matching and trade-off considerations.

The concluding chapter addresses three questions that are central to understanding migration in China in the twenty-first century. First, mobility increase continues unabated, and migrant work has firmly established itself as a way of life in many parts of the countryside. Second, in addition to circulation between the rural origin and urban place of work, peasant migrants are also circulating among places of migrant work, leaving undesirable jobs in pursuit of better returns and triggering labor shortages in southern China. Their strategies, however, have hitherto not manifested into systemic labor resistance. Third, the Hu Jintao–Wen Jiabao administration is paying greater attention to migrant issues such as back wages and social protection. Finally, I reiterate the argument that the hukou and permanent migrant paradigms are inadequate, and I argue that greater attention should be paid to persistent inequalities in Chinese society and to the circularity of migrants.

2 Volume and spatial patterns of migration

Introduction

Historically, population mobility in China has been low. This reflects partly the agrarian nature of the economy which bound people to the land. Low rates of rural–urban migration also explained the persistently low levels of urbanization. As late as 1953, the level of urbanization was only 13 percent (Zhou and Ma 2005). Over the past five decades, mobility levels and rates have fluctuated – they increased during the 1950s, decreased sharply during the 1960s due to migration control (see Chapter 3), and then increased in the 1970s and especially since the 1980s (Yang 1994: 103–122). This chapter aims at documenting the volume, rates and spatial patterns of migration between the mid-1980s and 2000, based primarily on comparisons between the 1990 and 2000 censuses. It begins by explaining migration as a flow measure and a stock measure and showing the trends of permanent and temporary migration. Discussion of the spatial patterns of migration focuses on places of origin and destination, interprovincial and interregional migration, and their changes.

Volume of migration

Despite the large number of studies on migration in China (see Chapter 1), there is still much confusion about the magnitude of migration. This is due in part to the existence of many different concepts and terms related to migration and the frequent changes of definition in census and census-type surveys in China (Duan and Sun 2006).[8] In the following, I describe the two most popular indicators – migration as a flow measure and floating population as a stock measure – and related concepts.

Migration as a flow measure

Migration volume is most commonly defined as the number of people who changed residence between two specific points in time. For example, the US census defines a person as a mover or migrant if his/her "usual place of residence" at the time of enumeration is different from that five years ago. More specifically, "movers" refers to all persons who have changed residence regardless of whether they have crossed a county boundary, whereas "migrants" refers to only those who have moved from one county to another. This definition excludes migrants who died between the two points in time, those who are younger than five years old at the time of enumeration, migrants who returned to the origin before the date of enumeration (i.e. return migration), and multiple moves that occurred between the two points in time (i.e. these are counted as one move). The overall effect of these limitations is underestimation of the true magnitude of migration. Despite these limitations, by standardizing the definition in both temporal and spatial terms, this approach facilitates comparison of mobility over time and across countries.

Similar approaches are used to measure migration volume in China. The 1990 census is the first census in China that includes systematic information about migration. According to that census, a two-step approach is used to define a migrant. First, a migrant must be five years or older on the date of enumeration (July 1, 1990) and whose usual place of residence is in a different county-level unit from that five years ago (July 1, 1985). This step is similar to the US census's definition of "migrants." The second step is concerned with an individual's hukou location or place of registration, a concept and practice tied to the hukou system (see also "Permanent migrants and temporary migrants" and Chapter 3). Specifically, a migrant must belong to one of the following two categories: (1) he/she is registered at the place of enumeration; (2) he/she has stayed at the place of enumeration for more than one year or has left his/her place of registration for more than one year. This temporal requirement results in underestimation of short-term migrants because individuals who have stayed at the place of enumeration for less than one year *and* who have left the place of registration for less than one year are counted at the place of registration rather than the place of enumeration and are not considered migrants (Banister and Harbaugh 1992).

In addition to moving the date of enumeration to November 1, the 2000 census introduces two changes to the definition of migration. First, the temporal requirement for category (2) is changed from one year to six months, thus counting also individuals who have stayed at the place of enumeration for more than six months and less than one year, or have left the place of registration for more than six months and less than one year. Though the exact effect of this change is difficult to determine, most researchers expect the effect to be relatively minor (Liang 2001a; Huadong shifan daxue 2005: 1216).

Second, the 2000 census counts as migrants not only persons who have moved between county-level units but also those who have moved within county-level units (between township-level units; see also "Spatial patterns of migration") (Wang and Ye 2004). In other words, the migration volume reported in the 1990 census includes only intercounty migrants (35.3 million) but the volume reported in the 2000 census includes not only intercounty migrants (79.1 million) but also intracounty migrants (42.2 million) (Table 2.1). The two total migration volumes – 35.3 million from the 1990 census and 121.2 million from the 2000 census – are, therefore, not comparable[9] (row 6 in Table 2.1). Comparison

Table 2.1 Migration volumes, 1990 and 2000 censuses

			1990 census		2000 census		2000/1990 ratio	
			Volume (million)	Rate (%)	Volume (million)	Rate (%)	Volume (million)	Rate (%)
Intraprovincial*	(1)	(2) + (3)	23.8	2.3	88.9	7.6	NA	NA
Intracounty	(2)		NA	NA	42.2	3.6	NA	NA
Intercounty	(3)		23.8	2.3	46.8	4.0	2.0	1.7
Interprovincial	(4)		11.5	1.1	32.3	2.8	2.8	2.5
Total intercounty	(5)	(3) + (4)	35.3	3.4	79.1	6.7	2.2	2.0
Intraprovincial share (%)		(3)/(5)	67.4		59.1			
Interprovincial share (%)		(4)/(5)	32.6		40.9			
Total*	(6)	(2) + (3) + (4)	35.3	3.4	121.2	10.3	NA	NA

Sources: 1990 census one percent sample; Liang and Ma (2004); NBS (2002: 1813–1817); Population Census Office (2002).

Notes
Rate: Percent of population aged 5+ at the census year.
* Not comparable between 1990 and 2000 censuses.

between the two censuses can, however, still be made for interprovincial migration (row 4), intercounty migration within provinces (row 3) and total intercounty migration (row 5). The 2.8 times increase of interprovincial migration from 11.5 million to 32.3 million and the 2.2 times increase of intercounty migration from 35.3 million to 79.1 million indicate that mobility increased considerably in the 1990s. The interprovincial and intercounty migration rates increased from, respectively, 1.1 percent and 3.4 percent in 1990 to 2.8 percent and 6.7 percent in 2000. Although these rates are still low compared to those of advanced industrial economies such as the US, the rapid increase of mobility in China during this period sets it apart from most other countries.[10] In addition, the share of interprovincial migrants among all intercounty migrants increased from 32.6 percent in 1990 to 40.9 percent in 2000, which suggests that the friction of distance has decreased over time (see also "Spatial patterns of migration").

The information on intracounty migration from the 2000 census is valuable for understanding mobility over relatively short distances, including movements from one urban district to another. In this book, in order to focus on migration that involves significant changes in terms of both distance and physical separation from the origin, I focus on intercounty (and interprovincial) migration rather than intracounty moves.

Permanent migrants and temporary migrants

As described earlier, definitions of migration in China involve the place of registration. The hukou system has, since the late 1950s, been used to define Chinese

22 Volume and spatial patterns of migration

citizens' place of registration. Details of the hukou system are the subject of Chapter 3. Suffice it to say that the place of registration (or hukou location) is considered to be where one belongs and where one is eligible for state-sponsored benefits (such as housing and health care in urban areas and access to farmland in rural areas). Until the mid-1980s, it was extremely difficult for rural Chinese to survive in cities because they did not have access to the necessities of life in cities, which were available only to urban residents (those who had urban hukou). The hukou system, therefore, kept rural–urban migration to a minimum. Since the mid-1980s, expanded options for rural Chinese to work in urban areas have fostered a surge in rural–urban migration (see Chapter 3). This is due to the relatively new phenomenon of *renhu fenli* – one's physical separation from one's place of registration. Specifically, rural Chinese are permitted to work in urban areas without obtaining urban hukou. Expanded markets for food and other necessities have made it possible for peasant migrants to live and work in cities. Nevertheless, from the perspective of the hukou system, they are temporary migrants because they do not have urban hukou (they have not "moved" the hukou location to the destination).

Internal migration in China since the 1980s is, therefore, aptly described as a two-track system consisting of permanent migrants and temporary migrants (Chan 1994; Gu 1992; Yang 1994). "Temporary migrants" refers to all migrants whose place of registration differs from the place of residence. It is the hukou status, rather than the duration of stay, that defines them as temporary migrants (Goldstein and Goldstein 1991). Most rural–urban migrants belong to this category. "Permanent migrants," in contrast, refers to migrants who have moved the hukou location to the destination, or put in another way, their place of registration is the same as their place of residence. Using the 1990 census definition described earlier, category (1) refers to permanent migrants and category (2) refers to temporary migrants. A variety of terminologies have been used to describe this dichotomy – hukou versus non-hukou migration, "plan" versus "non-plan" (or self-initiated) migration, formal versus informal migration, and de jure versus de facto migration (Chan *et al.* 1999; Fan 1999; Gu 1992; Li 1995; Yang 1994). However, the terms permanent migration and temporary migration are by far the most commonly used and are therefore the preferred terms in this book (e.g. Goldstein and Goldstein 1991; Goldstein and Guo 1992; Yang 2006; Yang and Guo 1996). In brief, permanent migrants are primarily those sponsored by the state and/or more skilled and highly educated, while temporary migrants are self-initiated, market-driven migrants and are mostly of lower socioeconomic statuses. In Chapter 4, I shall discuss in greater detail how permanent migrants and temporary migrants differ.

According to the 1990 census, respectively 19.1 million and 16.2 million, or 54.1 percent and 45.9 percent, of migrants are permanent migrants and temporary migrants (Table 2.2). In the late 1980s, therefore, permanent migration was still a dominant component of population movements in China. By the 2000 census, the volume of permanent migration remained around 20 million, but its share had declined sharply to only 25.6 percent. In contrast, the volume of temporary

Table 2.2 Permanent migrants, temporary migrants and floating population, 1990 and 2000 censuses

	1990 census		2000 census		2000/1990 ratio	
	Volume (million)	Rate (%)	Volume (million)	Rate (%)	Volume (million)	Rate (%)
Migrants (intercounty)						
Permanent migrants	19.1	1.8	20.2	1.7	1.1	0.9
Share (%)	54.1		25.6			
Temporary migrants	16.2	1.6	58.8	5.0	3.6	3.2
Share (%)	45.9		74.4			
Total	35.3	3.4	79.1	6.7	2.2	2.0
Floating population						
Intercounty	22.6	2.0	78.8	6.3	3.5	3.2
Intracounty	NA	NA	65.6	5.3	NA	NA
Total*	22.6	2.0	144.4	11.6	NA	NA

Sources: 1990 census one percent sample; Liang and Ma (2004); Population Census Office (2002).

Notes
Rate: For migrants, rate is calculated as percent of population aged 5+ at the census year; for floating population, rate is calculated as percent of total population at the census year.
* Not comparable between 1990 and 2000 censuses.

migration more than tripled from 16.2 million to 58.8 million and its share increased to 74.4 percent. These changes signal that, since the 1990s, self-initiated, market-driven migration has become the dominant form of mobility in China.

Floating population as a stock measure

Floating population is a unique concept in China and is tied to the hukou system. Individuals that are not living at their place of registration are considered "floating." Different from temporary migration, floating population is not based on comparing two points in time but is a stock measure that counts all individuals whose hukou location is not at the usual place of residence. A person who moved to the destination more than five years ago is still considered part of the floating population if he/she has not yet moved the hukou to the destination.

A temporal criterion usually qualifies the definition of floating population, and the criterion varies from one source to another.[11] The 1990 census specifies that members of the floating population must have left the hukou location or have lived at the usual place of residence for at least a year. In the 2000 census, the criterion is shortened to six months. The spatial criterion also changed between the two censuses, much like the change described earlier for the definition of migrants. In the 1990 census, the floating population includes persons who have moved from one county-level unit to another, whereas the 2000 census includes all persons who have moved across township-level units. In short, the 1990 census counts only intercounty floating population while the 2000 census counts both intercounty and intracounty floating population. The total volumes of the floating population as documented by the two censuses – 22.6 million in 1990 and 144.4 million in 2000 – are, therefore, not comparable (Table 2.2).[12] The volumes of intercounty floating population – comparable other than the change in temporal criterion from one year to six months – increased from 22.6 million or 2 percent of the population in 1990 to 78.8 million or 6.3 percent of the population in 2000. This rate of increase (3.5 times) is very similar to that of temporary migrants (3.6 times) between the two censuses (see Table 2.1).

Terminologies of migration

Not only are definitions of migration and migration-related measures, such as floating population, complex but the use of terminologies describing migration in China has been confusing and inconsistent (Duan and Sun 2006). In the Chinese-language literature, in particular, many terms have appeared and often the same term is used to refer to different things. The terms *qianyi* and *qianyi renkou* – best translated as, respectively, migration and migrant population – are sometimes used to refer to all migrants, much like the definitions described earlier in this chapter. Many researchers, however, use the terms strictly for permanent migration and permanent migrants (Yan 1998). In this tradition, terms such as *liudong renkou* (literally, floating population) and *wailai renkou* (population from outside) are used to refer to individuals who have not obtained hukou at the destination. However, the exact definitions – such as the temporal and spatial criteria – are often not clearly spelt out. The terms are sometimes used to refer to temporary migration as a flow measure and sometimes used to refer to floating population as a stock measure. And, some researchers combine *qianyi* and *liudong* into *liuqian*, in order to include all persons who have moved, regardless of their hukou status. Other terms will not be introduced here, but suffice it to say that much caution is needed when interpreting terms and statistics concerning migration in China. In this book, I rely heavily on the 1990 and 2000 censuses and focus specifically on migration as a flow measure, including permanent migration and temporary migration. I also draw upon data, material, and literature that focus on temporary migrants or floating population. The Sichuan and Anhui surveys (see Chapter 1), for example, include mainly peasants who are temporary migrants.

Spatial patterns of migration

Places of origin and destination

China has experienced more than two decades of relatively rapid urbanization (see Chapter 1).[13] Rural–urban migration has been the main source of urban growth (Chan and Hu 2003; Duan 2003; Lu and Wang 2006).[14] Estimation of the volume of rural–urban migration, however, is extremely difficult, not only because the definitions of urban population and urban places are complex and have frequently changed but also because available data, including census data, are mostly collected based on the spatial administrative hierarchy rather than urban and rural definitions.

A full-scale discussion of China's spatial administration is beyond the scope of this book. Briefly, China's spatial administration is most commonly defined in terms of four levels: provincial (*shengji*),[15] prefecture (*diji*),[16] county (*xianji*) and township (*xiangji*), in descending order of the hierarchy. The county and township levels are most relevant to the understanding of migration data as recorded in the censuses. There are two types of county-level (third-level) units – "cities" and "counties."[17] Although cities as a whole are more urbanized than counties, the former may contain significant numbers of rural population (Zhou and Ma 2005) and the latter may contain significant numbers of urban population. At the fourth, or township, level, there are three types of units: "streets" (*jiedao*), "towns" (*zhen*), and "townships" (*xiang*).[18] In the urban–rural continuum, streets are the most urbanized and townships are the most rural. Streets are further subdivided into "residents' committees" (*jumin weiyuanhui*); towns are subdivided into "residents' committees" and "villagers' committees" (*cunmin weiyuanhui*); and townships are subdivided into "villagers' committees." Since about 1983, residents' committees and villagers' committees have often been used to estimate, respectively, urban and rural population (Chan 1994: 22).

Both the 1990 and 2000 censuses record for migrants their place of residence at the time of enumeration as well as the place from which they migrated. The former is the migrant's destination and the latter is his/her origin. Interpretation of origins and destinations is, however, complicated by two issues. First, origins are recorded at the township level (streets, towns, townships) while destinations are recorded at the county level (cities, counties). Second, the origin and destination categories differ between the two censuses (Table 2.3). In the 1990 census, towns as origins are regarded as one category, but in the 2000 census, towns are further subdivided into residents' committees and villagers' committees. In addition, the 2000 census adds towns as a destination category, in addition to cities and counties. There is little documentation about the rationale of these changes; neither is there information about whether the town destinations refer to only towns administered by counties or also towns administered by cities (see note 13). A plausible reason for the increased attention on towns is that a large number of townships were "upgraded" to towns since the mid-1980s and especially in the 1990s, thus increasing tremendously the number of towns as well as town population (Zhou and Ma 2003). In the absence of detailed information, however, interpretation of origin and destination must be done with caution.

Based on the 1990 census, townships are the most popular origins (61.5 percent) of intercounty migrants, followed by streets (19.6 percent) and towns (19 percent). Cities attract the majority – 56.6 percent – of intercounty migrants and are more popular destinations than counties (43.4 percent). Township–city moves constitute the largest proportion (33.1 percent) of all combinations, followed by township–county moves (28.4 percent). Among interprovincial migrants, townships are again the most popular origins (59.8 percent) and cities are slightly

Table 2.3 Origin and destination types, 1990 and 2000 censuses

Origin	Destination												
	1990 census						2000 census						
	Intercounty migrants			Interprovincial migrants			Interprovincial migrants						
	Cities	Counties	Total	Cities	Counties	Total		Cities	Towns	Counties	Total		
Streets	12.2	7.4	19.6	17.7	8.6	26.3		9.2	1.6	2.0	12.8		
Towns	11.3	7.7	18.9	6.4	7.5	13.9		24.7	10.6	12.6	48.0		
(residents' committees)	–	–	–	–	–	–		(4.4)	(1.5)	(1.3)	(7.2)		
(villagers' committees)	–	–	–	–	–	–		(20.3)	(9.1)	(11.3)	(40.8)		
Townships	33.1	28.4	61.5	26.7	33.1	59.8		19.4	8.2	11.7	39.2		
Total	56.6	43.4	100.0	50.8	49.2	100.0		53.4	20.4	26.2	100.0		

Sources: 1990 census one percent sample; 2000 census 0.1 percent interprovincial migrant sample.

more popular (50.8 percent) than counties (49.2 percent) as destinations. Township–city moves constitute the second largest proportion (26.7 percent), after township–county moves (33.1 percent), of all combinations. Despite the impreciseness of the origin and destination categories to represent urban and rural places, the data do suggest that the bulk of the migrants came from rural origins and that urban destinations are more popular than rural destinations. It is important to note, however, that township–county moves do account for significant proportions of migrants, especially among interprovincial migrants, suggesting that rural–rural migration was an important component of migration in the late 1980s.

According to the 2000 census, towns (48 percent) have overtaken townships (39.2 percent) as the most popular origins of interprovincial migrants. However, this may simply reflect the increased number of towns during the 1990s, as described earlier. In fact, within towns, villagers' committees are much more popular origins (40.8 percent) than residents' committees (7.2 percent). If one considers townships and villagers' committees of towns as rural places, then 80 percent of interprovincial migrants have rural origins. Although the 1990 census data do not permit disaggregating towns into residents' and villagers' committees, the sum of towns and townships in that census accounts for only 73.7 percent of interprovincial migrants. These data, therefore, strongly suggest that a greater proportion of interprovincial migrants in the late 1990s than that in the late 1980s came from rural origins.

The data also show that an increased share of migrants is moving to urban destinations. According to the 2000 census, the proportion of interprovincial migrants moving to cities is 53.4 percent and is bigger than the 1990 census counterpart (50.8 percent). Moreover, if one considers towns, or some parts of towns, as also urban, then the increase between the two censuses would be even greater.

Putting origins and destinations together, and considering townships and villagers' committees as rural places and cities and towns as urban places, rural–urban migration accounts for 57 percent of all interprovincial migration, according to the 2000 census. More conservatively, if only cities are considered urban (and towns are not), then the proportion is 39.7 percent. Both are larger than the township–city share (26.7 percent) and the share of townships and towns combined to city destinations (33.1 percent) according to the 1990 census. Despite the constraints and crudeness of the data, therefore, it is reasonable to conclude that the share of rural–urban migrants among all migrants increased between the 1990 and 2000 censuses.

Interprovincial migration

As described earlier, by the time of the 2000 census, the share of interprovincial migration in total intercounty migration had increased to more than 40 percent. In addition, as I shall show below, the share of migration across the eastern, central and western regions had also increased considerably (Figure 2.1). These three regions – also referred to as the "three economic belts" – reflect a conceptualization popularized by the Seventh Five-Year Plan (1985–1990) that each of the regions

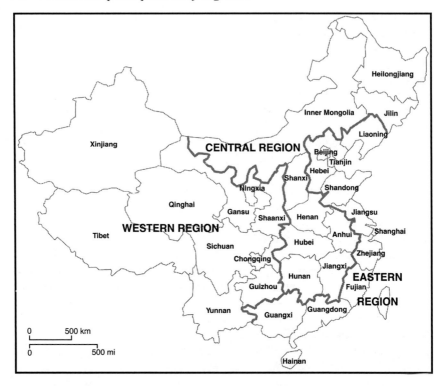

Figure 2.1 Provincial units and the three economic belts.

should focus on its comparative advantage (Fan 1995). They also constitute a convenient regionalization scheme to describe the level and changes of regional inequality, which are important for understanding the spatial patterns of migration.

Table 2.4 shows for all the provinces and regions their in-migration, out-migration and net migration according to the two censuses and their population and GDP per capita for 1988 and 1998 – the two mid-points of the census periods – and 2004. The migration data further confirm a considerable increase in interprovincial migration between the two censuses. Among the three regions, the eastern region accounted for about 41 percent of China's population but respectively 57 percent and 78.4 percent of interprovincial in-migration in the two periods. This region has been, no doubt, a gainer in migration, as further evidenced by its highly positive net migration figures. Conversely, the central and western regions as a whole lost population due to migration; that is, they had negative net migration in both periods. Moreover, the discrepancy in net migration between the eastern region on one hand and the central and western regions on the other increased, as did the discrepancy between the major donor provinces and receiving provinces, between the two censuses. This shows that gainers are gaining more and losers are losing more and that interprovincial

Table 2.4 Interprovincial migration, GDP per capita and population, 1990 and 2000 censuses

Province	Interprovincial migration ('000)							GDP per capita (2004 constant yuan)			Average annual growth (%)		Population (million)		
	1990 census			2000 census											
	In	Out	Net	In	Out	Net		1988	1998	2004	1988–1998	1998–2004	1988	1998	2004
Eastern	6,129	4,174	1,955	24,945	7,807	17,138		3,425	9,034	18,217	9.7	11.7	449.6	507.4	541.3
% of nation	57.0	38.8		78.4	24.5								41.4	41.2	41.9
Beijing	663	123	539	1,888	174	1,714		8,556	14,942	28,689	5.6	10.9	10.8	12.5	14.9
Tianjin	312	86	225	491	104	387		5,803	12,652	28,632	7.8	13.6	8.4	9.6	10.2
Hebei	468	665	−197	769	872	−103		2,386	6,431	12,878	9.9	11.6	58.0	65.7	68.1
Shandong	609	523	86	903	878	26		2,637	7,892	16,874	11.0	12.7	80.6	88.4	91.8
Liaoning	517	272	245	754	380	375		4,270	8,615	16,297	7.0	10.6	38.2	41.6	42.2
Shanghai	655	150	505	2,167	163	2,004		10,523	22,981	42,768	7.8	10.4	12.6	14.6	17.4
Jiangsu	837	588	248	1,907	1,240	667		3,710	9,505	20,723	9.4	13.0	64.4	71.8	74.3
Zhejiang	321	626	−305	2,714	968	1,746		3,648	10,724	23,820	10.8	13.3	41.7	44.6	47.2
Fujian	294	228	67	1,346	624	722		2,448	9,443	17,241	13.5	10.0	28.5	33.0	35.1
Guangdong	1,162	250	911	11,500	438	11,062		3,528	10,717	19,315	11.1	9.8	59.3	71.4	83.0
Hainan	133	112	22	218	130	88		2,737	5,616	9,405	7.2	8.6	6.3	7.5	8.2
Guangxi	157	549	−391	287	1,838	−1,551		1,577	3,897	6,791	9.0	9.3	40.9	46.8	48.9
Central	2,814	3,690	−876	3,246	15,590	−12,344		2,327	5,117	9,481	7.9	10.3	388.4	440.3	454.3
% of nation	26.1	34.3		10.2	49.0								35.7	35.8	35.2
Anhui	343	538	−195	313	2,892	−2,579		2,045	4,414	7,449	7.7	8.7	53.8	61.8	64.6
Shanxi	269	227	42	382	333	49		2,402	4,791	9,123	6.9	10.7	27.6	31.7	33.4
Inner Mongolia	239	278	−39	325	441	−116		2,431	4,992	11,376	7.2	13.7	20.9	23.5	23.8
Jiangxi	226	277	−52	235	2,680	−2,445		1,893	4,283	8,160	8.2	10.7	36.1	41.9	42.8

continued

Table 2.4 continued

Province	Interprovincial migration ('000)						GDP per capita (2004 constant yuan)			Average annual growth (%)		Population (million)		
	1990 census			2000 census										
	In	Out	Net	In	Out	Net	1988	1998	2004	1988–1998	1998–2004	1988	1998	2004
Jilin	254	346	–92	254	529	–275	3,159	5,803	10,920	6.1	10.5	23.7	26.4	27.1
Heilongjiang	332	594	–262	301	940	–639	3,301	7,153	13,893	7.7	11.1	34.7	37.7	38.2
Henan	493	578	–85	468	2,306	–1,838	1,809	4,698	9,072	9.5	11.0	80.9	93.2	97.2
Hubei	411	348	62	605	2,209	–1,604	2,684	5,962	10,489	8.0	9.4	51.9	59.1	60.2
Hunan	248	504	–256	362	3,260	–2,898	2,269	4,903	8,379	7.7	8.9	58.9	65.0	67.0
Western	1,818	2,897	–1,078	3,621	8,415	–4,794	2,027	3,916	7,215	6.6	10.2	248.7	282.6	295.9
% of nation	16.9	26.9	–	11.4	26.5	–	–	–	–	–	–	22.9	23.0	22.9
Sichuan	410	1,287	–877	660	5,091	–4,432	2,005	4,194	7,784	7.4	10.3	105.8	115.5	118.5
Guizhou	198	309	–111	261	1,231	–971	1,470	2,210	4,078	4.1	10.2	31.3	36.6	39.0
Yunnan	232	272	–40	731	397	334	1,724	4,194	6,703	8.9	7.8	35.9	41.4	44.2
Shaanxi	301	332	–31	420	716	–296	2,255	3,705	7,783	5.0	12.4	31.4	36.0	37.1
Gansu	159	269	–110	203	555	–353	2,015	3,352	5,952	5.1	9.6	21.4	25.2	26.2
Qinghai	104	98	5	76	120	–44	3,137	4,380	8,641	3.3	11.3	4.3	5.0	5.4
Ningxia	78	56	22	129	87	41	2,398	4,074	7,829	5.3	10.9	4.5	5.4	5.9
Xinjiang	336	273	63	1,142	216	926	3,230	6,061	11,208	6.3	10.2	14.3	17.5	19.6
All provinces	10,761	10,761	0	31,813	31,813	0	2,712	6,457	13,167	8.7	11.9	1,086.8	1,230.3	1,291.4

Sources: Compiled by author from China Statistical Yearbooks of various years; Hsueh *et al.* (1993); NBS (1999; 2002: 1813–1817); and SSB (1992: 126–139).

Note
Tibet is not included and because of data constraints Sichuan is combined with Chongqing.

migration is playing an increasingly important role in redistributing population between provinces (Fan 2005a; 2005b).

Data for GDP per capita – a commonly used proxy for the level of economic development – show that the eastern region has had the most rapid economic growth and that the gap between it and the central and western regions has widened over time. The eastern region grew at average annual rates of 9.7 percent and 11.7 percent respectively during the 1988–1998 and 1998–2004 periods, whereas the average growth rates of the central and western regions were in the order of 6 percent and 7 percent in the first period and 10 percent in the second period. In 2004, GDP per capita for the eastern region as a whole stood at 18,271 yuan, nearly twice that of the central region and two and a half times that of the western region. The gap between the central and western regions has also widened but to a smaller degree.

At the provincial level, the three centrally administered municipalities – Shanghai, Beijing and Tianjin – were the leaders in GDP per capita throughout the period 1988 to 2004. Most other eastern-region provinces, the northeastern provinces of Heilongjiang and Jilin, and the northwestern provinces of Xinjiang had higher levels of GDP per capita than the rest of the country. Though the general pattern of regional disparity persisted between 1988 and 2004, the absolute gaps among provinces grew wider. The contiguous coastal region extending from Beijing to Guangdong – comprising in particular Beijing, Tianjin, Jiangsu, Shanghai, Zhejiang, Fujian and Guangdong – has emerged as the growth core of China since the 1980s (Fan 1995). The prominence of Xinjiang in the west is also notable, due in no small part to robust cross-border trade between China and the Central Asian Republics (Loughlin and Pannell 2001). Much of central and western China, however, remains poor. The gap between Shanghai and Guizhou, respectively the highest and lowest-ranked provinces in terms of GDP per capita, widened from 9,053 yuan in 1988 to 38,691 yuan in 2004. Accordingly, the Shanghai–Guizhou ratio surged from 7.2:1 to 10.5:1, depicting an increase in not only the absolute level of disparity but also relative disparity. The disparity in economic development among the three regions and at the provincial level correlates with the spatial patterns of interprovincial migration (Fan 2005a; 2005b). Provinces with the largest in-migration, including Guangdong, Zhejiang, Shanghai, and Beijing, are also among the top in GDP per capita. Conversely, provinces with the largest out-migration, such as Sichuan, Hunan, Anhui, Jiangxi, and Henan, have relatively low GDP per capita.

In addition, population size is related to the size of interprovincial migration. Most of the leading donor provinces (as indicated by large volumes of out-migration) have large population. For example, Sichuan was the largest donor of interprovincial migration in both periods and was the most populous province in China. Other major donor provinces with large population include Henan, Hebei, Hunan, and Anhui. Among prominent receiving provinces (as indicated by large volumes of in-migration), Shandong, Guangdong, and Jiangsu are respectively the third, fourth, and fifth most populous provinces in China. There are, nevertheless, important exceptions to the relationship between population size and in-migration. Specifically, the municipalities of Beijing, Tianjin, and

Shanghai, which are densely populated but have relatively small population compared to other provincial units, are among the leading receiving provinces, underscoring the attractiveness of large, coastal cities to migrants.

Interregional migration

To more specifically identify migration flows among the three regions, Table 2.5 shows the proportions of interprovincial migration attributable to intraregional and interregional flows, represented by respectively diagonal and off-diagonal cells. In both periods, the eastern region had the largest diagonal proportion – 24.4 percent and 18.4 percent, indicating more active interprovincial migration there than within the other two regions (Cai and Wang 2003). Intraregional flows, however, declined in relative importance. Between the two censuses, the sum of off-diagonal proportions increased from 57.3 percent to 71.8 percent, depicting the increased prominence of interregional flows relative to intraregional flows. Of the six off-diagonal cells, only two – the central-to-eastern and the western-to-eastern – increased over time, indicating an acceleration of migration flows from the two non-coastal regions to the eastern region. The flow from the central region to the eastern region is especially noteworthy, as its share almost doubled, from 21 percent to 41.8 percent. The western-to-central flow (6.3 percent) is bigger than that of central-to-western flow (4.1 percent), according to the 1990 census, but the latter (3.2 percent) is bigger than the former (2.4 percent) according to the 2000 census. This change depicts the increased prominence of western-region provinces such as Xinjiang and Yunnan in attracting migrants.

Figure 2.2 illustrates further the volumes of interprovincial migration within

Table 2.5 Proportions of interprovincial migration within and among regions, 1990 and 2000 censuses

Origin	Destination			
	Eastern	Central	Western	Total
1990 census				
Eastern	24.4	10.7	3.7	38.8
Central	21.0	9.2	4.1	34.3
Western	11.5	6.3	9.1	26.9
Total	57.0	26.1	16.9	100.0
2000 census				
Eastern	18.4	3.8	2.4	24.5
Central	41.8	4.0	3.2	49.0
Western	18.2	2.4	5.8	26.5
Total	78.4	10.2	11.4	100.0

Sources: NBS (2002) and SSB (1992).

Note
See Table 2.4.

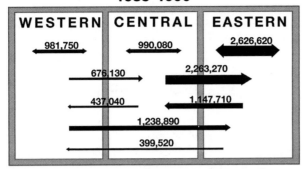

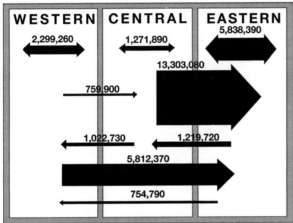

Figure 2.2 Volumes of interprovincial migration within and among regions, 1990 and 2000 censuses (sources: NBS (2002); SSBC (1992); reproduced from Fan (2005a) with permission).

Note
See Table 2.4.

and among the three regions. Between the 1990 and 2000 censuses, all flows had increased in volume, with the most pronounced increases within the eastern region and those from the central and western regions to the eastern region. The 1995–2000 central-to-eastern flow is six times greater than the 1985–1990 flow and the western-to-eastern flow stands five times greater.

Interprovincial net migration

By mapping the net migration rates of provinces, Figures 2.3 and 2.4 seek to identify more specifically the gainers and losers of migration at the provincial level. Based on the 1990 census, Beijing and Shanghai have the highest rates. Other eastern-region provinces including Liaoning, Tianjin, Shandong, Jiangsu, Fujian, Guangdong and Hainan, the two central-region provinces of Shanxi and Hubei, and the three western-region provinces of Ningxia, Qinghai and Xinjiang, also demonstrate positive migration rates. The rest of the country have negative rates. By the 2000 census, variations in provincial rates have further increased. Guangdong, Beijing and Shanghai lead the nation with two-digit positive rates. Other eastern-region provinces, except Hebei and Guangxi, all register positive net migration rates. As in the previous period, Shanxi, Ningxia and Xinjiang continue to have positive rates, and they are further joined by Yunnan in the southwest. Xinjiang, in particular, produces a larger net migration rate than most other provinces, suggesting that its economic growth related to cross-border trade is a significant attraction to migrants (Loughlin and Pannell 2001). Yunnan's case is less clear but its attraction to migrants possibly reflects its success in tobacco production (Shen 2001). Both Xinjiang and Yunnan are members of the western region and their positive rates have likely contributed to the reversal of positions between the central and western regions, as observed earlier (Table 2.5 and Figure 2.2).

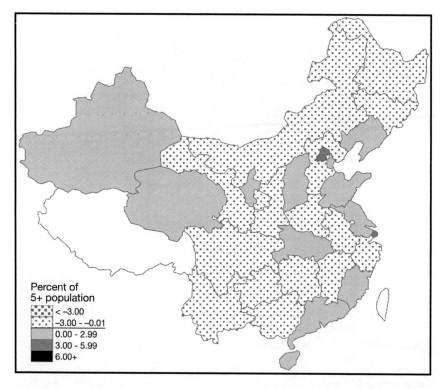

Figure 2.3 Interprovincial net migration, 1990 census (sources: SSBC (1992); adapted from Fan (2005a) with permission).

Note
See Table 2.4.

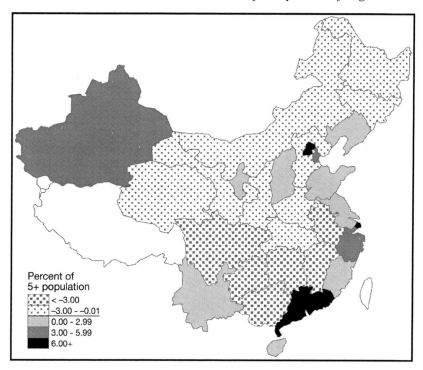

Figure 2.4 Interprovincial net migration, 2000 census (sources: NBS (2002); adapted from Fan (2005a) with permission).

Note
See Table 2.4.

Among provinces with negative net migration rates, the most prominent ones – with rates more negative than –3 – constitute a contiguous zone spanning south central and southwestern China and including Anhui, Jiangxi, Hunan, Guangxi, Guizhou, and Sichuan. Sichuan is the largest source of interprovincial migration (Table 2.4). As shown earlier and in Table 2.4, these provinces are among the least developed in China. Recent research has documented that Anhui, Jiangxi, Hunan, and Sichuan suffered from negative employment growth in the 1990s (Yang *et al.* 2003). And Anhui, Jiangxi and Hunan – all in the central region – have the most negative net migration rates, in the order of –5 percent to –7 percent. Clearly, these three provinces have contributed significantly to making the central region the largest donor of migrants among the three regions.

Interprovincial migration flows

Figures 2.5 and 2.6 illustrate the 30 largest interprovincial migration flows, which account for respectively 31.5 percent and 56 percent of the total interprovincial migration, according to the 1990 and 2000 censuses. The

36 *Volume and spatial patterns of migration*

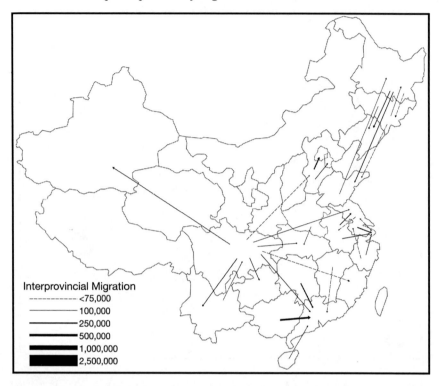

Figure 2.5 The 30 largest interprovincial migration flows, 1990 census (sources: SSB (1992); adapted from Fan (2005b) with permission).

Note
See Table 2.4.

figures depict the most prominent patterns of interprovincial migration and their changes between the two censuses. First, they support the observation that migration is related to the population size of origin and destination provinces. Sichuan as a leading origin province and Guangdong as a leading destination province are most prominently illustrated. Second, the figures show that migration flows are a function of distance. Although long-distance moves such as those from Sichuan to Xinjiang and from Heilongjiang to Shandong are among the 30 largest flows, migration between adjacent provinces is more numerous and voluminous. In fact, several clusters of proximate provinces with large migration flows among them resemble regional migration fields (He and Pooler 2002). The 1990 census identifies at least four clusters: among the northeastern provinces of Heilongjiang, Jilin and Liaoning and between them and Shandong; among Jiangsu, Shanghai, Zhejiang, and Anhui; from Jiangxi, Hunan, Guangxi, and Hainan to Guangdong; from Sichuan to Hubei and Guizhou and between Sichuan and Yunnan. Some of the migration fields are attributable to past migration patterns. The flows between Heilongjiang

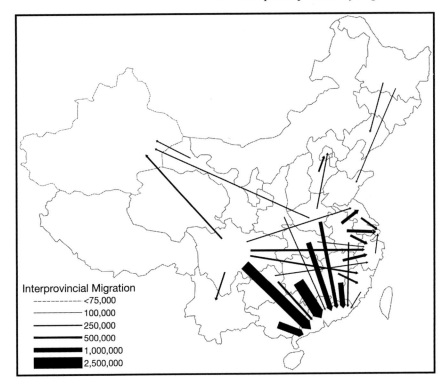

Figure 2.6 The 30 largest interprovincial migration flows, 2000 census (sources: NBS (2002); adapted from Fan (2005b) with permission).

Note
See Table 2.4.

and Shandong, for example, reflect in part interactions that began during the pre-reform rustication programs and the return of migrants from frontier regions to their coastal origins (Chan *et al.* 1999). By the 2000 census, most of the migration fields persist, depicting the continued negative effect of distance. However, the relative prominence of some long-distance migration flows, such as that from Henan to Xinjiang, and the strong pull of Guangdong to migrants from a large number of provinces, adjacent or otherwise, suggest that the friction of distance has decreased over time.[19]

Third, the spatial patterns clearly point to the effect of uneven regional development. Most of the prominent origin provinces are relatively poor and the most prominent destination provinces are economically more developed. The largest migration flows are toward the eastern region. The only noticeable exception is Xinjiang, which has enjoyed near-average rates of economic growth as a result of its cotton industry and cross-border trade with the Central Asian Republics (Liang and Ma 2004; Loughlin and Pannell 2001; Pannell and Ma 1997). Changes between the two periods further reinforce the notion that economic development is an increasingly important factor of migration. Figure 2.5 shows that in the late 1980s a number of migration streams are accompanied by counter-streams, including those between Heilongjiang and Jilin, Heilongjiang and Shandong, Jiangsu and Shanghai, Jiangsu and Anhui, and Sichuan and Yunnan. By contrast, in the late 1990s all the 30 largest migration flows are unidirectional. In other words, counter-streams are considerably smaller in volume and are thus not among the 30 largest flows. This suggests that over time interprovincial migration has become more concentrated and more "efficient" in redistributing population (He and Pooler 2002; Fan 2005a). In addition, the 30 largest flows account for a significantly greater proportion of the total migration volume in the 2000 census (56.0 percent compared to 31.5 percent in the 1990 census). Thus, over time, interprovincial migration has become more efficient, more unidirectional and more concentrated. The eastern coastal provinces of Jiangsu, Shanghai, Zhejiang, Fujian, and Guangdong, in particular, have clearly become the most dominant destinations. The pull of Guangdong is especially profound, attracting migrants from both nearby and distant provinces. According to the 2000 census, Guangdong alone accounts for 36 percent (11.5 million) of all interprovincial in-migrants in China. The changes between the two periods suggest that the most developed parts of the country are increasingly attractive to migrants and that economic development has become an increasingly powerful predictor of migration flows.

Despite changes between the two periods, the largest migration flows display a high degree of consistency. Evidence includes the persistence of regional migration fields, as described earlier. In addition, half of the 30 largest flows in the 1990 census are also among the 30 largest in the 2000 census. Some of the migration streams have persisted and gained magnitude over time. They include, for example, from Sichuan to Guangdong, Fujian, Jiangsu, Xinjiang, and Yunnan; from Anhui to Jiangsu and Shanghai; from Guangxi, Hunan, and Jiangxi to Guangdong; and from Hebei to Beijing. While these flows are attributable in part to the effects of population size, distance, and economic development, their persistence and increased magnitude over time suggest that present migration is influenced by past migration. The persistence of migration streams reflects the role of networks – commonly measured by migrant stock – which facilitate migration and channel migrants from specific origins to specific destinations.

The evidence shown above suggests that population size, distance, economic development, and migrant stock are prominent factors explaining the size of interprovincial migration. Using the gravity model as a basis and including the above factors, I have shown elsewhere (Fan 2005b) that these are indeed powerful predictors of interprovincial migration flows in China. In addition, statistical analysis has shown that migrant stock is the most important determinant of interprovincial migration, and that over time the destination's economic development has increased in importance and distance has decreased in importance. These results shed light on processes and the decision-making considerations of migrants; namely, they rely heavily on networks in choosing among destinations, they are increasingly responsive to economic opportunities in the destination, and they are increasingly capable of overcoming the friction of distance.

Summary and conclusion

Internal migration in China, whether evaluated by flow or stock measures, has increased manyfold since the 1980s. Between the 1990 and 2000 censuses, intercounty migration has more than doubled in size, interprovincial migration has almost tripled in size, and intercounty floating population has increased by three and a half times. Temporary migrants account for less than half of all intercounty migrants according to the 1990 census, but have increased their share to almost three-quarters by the 2000 census. Thus, not only has there been a dramatic surge of migration but population movements are also increasingly defined by self-initiated, market-driven processes that characterize temporary migrants (see Chapter 4).

Comparing the two censuses, rural origins and urban destinations account for increasing shares of migrants. Rural-to-urban moves, therefore, are increasingly the dominant mode of population movements in China. The share of interprovincial migration, which as a whole entails longer distances than intraprovincial migration, has increased from less than two-thirds to more than 40 percent between the 1990 and 2000 censuses. Among interprovincial flows, the share of interregional migration – moves among the eastern, central, and western regions – as opposed to moves within the respective regions, has increased by almost 15 percent point to about 72 percent by the 2000 census. These findings show that the friction of distance has declined. Analysis of interprovincial migration indicates that population movements are increasingly unidirectional and concentrated. Over time, sending provinces are losing more population and destination provinces are gaining more population because of migration. Provinces in south central and southwestern China are the largest donors of migrants, whereas those along the eastern coast, especially Guangdong, have become major magnets to migrants. The above spatial patterns underscore the strong relationship between regional economic development and migration – sending provinces are primarily less developed and receiving provinces are overwhelmingly those with higher levels of economic development and promising employment opportunities. This is strong evidence that the pursuit of economic betterment is the primary objective of migrants. The persistence of migration streams also highlights the role of networks among migrants.

The volume and spatial patterns of migration, as illustrated in this chapter, directly inform this book's arguments about the state and the household. The increased magnitude and importance of temporary migrants, most of whom are rural–urban migrants, reinforces the centrality of the hukou system and the peasant household for understanding mobility in China. Chapter 3 examines in detail the hukou system and its implications for temporary migration and Chapter 4 focuses on types and processes of migration in relation to hukou. Chapter 5 and other parts of the book address considerations of peasant migrants and their households, further elaborating the importance of economic motives and networks in their migration strategies.

3 The hukou (household registration) system

> After birth you should get a hukou right away. You need hukou to enter kindergarten; and you need local hukou to find a job. When you date, you should know the other person's hukou. All kinds of permits can only be processed with hukou; and all kinds of benefits depend on your hukou. When you move to another place, you need to change the hukou. When you die, remove your hukou.
>
> Depeng Yu (2002: 12)

Introduction

The hukou system was formally implemented in the late 1950s. Over the past several decades, it has not only fortified institutional and social barriers between rural and urban China but also influenced all aspects of Chinese society and economy. In the Maoist, pre-reform period, these barriers enabled the state to pursue development plans by binding peasants to the countryside and fostering the transfer of value from agriculture to industry. Despite the economic reforms and many changes to hukou regulations that have facilitated rural–urban migration since the 1980s, formidable barriers between the rural and the urban segments of society persist. These barriers, as I have argued in Chapter 1, have enabled the state's pursuit of developmentalist goals in a new milieu open to the market and the global economy. The hukou system, in both the pre-reform and reform periods, has been a crucial tool of the state and has had deep impacts on the lives of Chinese citizens. This chapter focuses on the hukou system, its origins, its role in migration control, criticisms of the system, and hukou reforms.

Hukou type and hukou location

Hukou is a form of population registration formally required and legalized since the National People's Congress (NPC) promulgated the People's Republic of China Regulations on Household Registration (*Zhongguo renmin gongheguo hukou dengji tiaoli*) on January 9, 1958 (Chan and Zhang 1999; Cheng and Selden 1994; *China Daily* 2001; Yu 2002: 18). Under the regulations, every Chinese citizen must be registered and the registration must be under one unit and one unit only (in one place and one place only). The most common unit for the registration is the household, which maintains a "hukou registration book" that records who belong to the household and their hukou types. In addition, a very small number of persons are registered under "collective units." University students, for example, are expected to temporarily shift their registration to the university while they are enrolled.

The term hukou refers specifically to two aspects of the registration system. The first aspect is hukou type or classification (hukou *leibie*) and consists of the "agricultural" (*nongcun*) and "non-agricultural" (*feinong*) categories (Chan and Zhang 1999; *China Daily* 2001). When the categories first came into being, they reflected closely where people lived and to a certain extent their occupations, because agricultural hukou was assigned to residents living in the countryside and non-agricultural hukou was assigned to those living in urban areas. These categories continue to distinguish rural Chinese from urban Chinese.[20] However, the occupational and geographical meanings of the two adjectives "agricultural" and "non-agricultural" are increasingly diluted, as many rural residents and rural–urban migrants engaged in urban work still have agricultural hukou. More accurately, therefore, agricultural and non-agricultural hukou refer to, respectively, the rural and urban statuses of individuals. Those with non-agricultural hukou are entitled to state welfare, benefits, and subsidies which during the pre-reform period were designed to take care of the individual from cradle to grave and which today continue to privilege urban Chinese over their rural counterparts. In contrast, in both the pre-reform and reform periods, Chinese citizens with agricultural hukou received little state support other than the right to farm.

The second aspect is hukou location, also referred to as hukou place or place of registration. The Chinese term for hukou location – hukou *suozaidi* – is literally translated as "where the hukou resides." As described earlier, for the vast majority of Chinese, their hukou is associated with a household and hence their hukou location is defined by the location of the household. Hukou location enables the individual to gain access to benefits in a specific locality that are not normally available to individuals whose hukou location is elsewhere. In practical terms, having an urban area as the hukou location is advantageous because of the abundance of state-sponsored benefits in urban areas compared to rural areas, and hukou location in large cities is superior to that in small cities and towns. For example, having a Beijing hukou opens the door to certain jobs closed to non-Beijing residents. Among rural areas, a hukou location in prosperous areas is more attractive than that in less prosperous areas. For example, a rural woman married to a man in a prosperous village would be eager to have her hukou moved to the husband's village in order to gain access to the farmland there. In the eye of the state, hukou location is where one belongs (Wang, Fei-Ling 2005: 10). For example, even though a rural person may have stayed and worked in the city for an extended period of time, he/she is still regarded as a temporary migrant rather than a permanent urban resident so long as he/she does not have "local hukou" (*bendi* hukou) (see also Chapter 2). Institutionally, the rural person belongs to the village where his/her hukou resides, and is considered a permanent resident there even though he/she may have left the origin many years ago. Stated differently, a rural migrant, after working in a city for years, will not be eligible for urban benefits or services in the absence of that city's hukou and, indeed, will remain vulnerable to being ejected from the city.

The assignment of hukou type and hukou location is primarily by birth and is inherited from one generation to the next. For most people, therefore, hukou is an ascribed status (Wu and Treiman 2004; Yu 2002: 52). Changing the hukou type or hukou location requires approval from state authorities. The conversion process from agricultural hukou to non-agricultural hukou, referred to as *nongzhuangfei*, is complex and strictly controlled. It is subject to both "policy" (*zhengce*), which defines the qualifications of individuals eligible to apply for non-agricultural hukou, and "quota" (*zhibiao*), which is set by the State Planning Commission and limits the number of people that may be granted non-agricultural hukou (Chan and Zhang 1999). Enrollment in specialized secondary schools or higher education institutions in urban areas, recruitment by state-owned enterprises (SOEs), and promotion to a senior administrative post – all involving a small and upwardly mobile minority of the population – are some of the most common means by which non-agricultural hukou may be obtained (e.g. Yu 2002: 52). In these cases, non-agricultural hukou serves as a reward to the individuals' special achievements. In addition, the state may also grant non-agricultural hukou for political and institutional reasons. For example, the return of demobilized military personnel and of individuals who had been "sent down" to the countryside during the 1960s and 1970s rustication movement may be accompanied by the granting of non-agricultural hukou. Wu and Treiman (2004) show that joining the People's Liberation Army (PLA) and to a certain extent having Communist Party membership increase the chance of conversion from agricultural hukou to non-agricultural hukou. Cadres of county and township governments at senior levels may be considered urban representatives of the state and awarded non-agricultural hukou (Knight and Song 1999: 9). The above cases are largely regulated by central-planning mechanisms and are thus considered "within state plan" arrangements. Overall, hukou conversion from agricultural to non-agricultural is only available to a very small proportion of the rural population, and the rate of conversion remained low throughout the 1980s and 1990s (Wu and Treiman 2004).

Changing the hukou location from one place to another requires approval by the public security authorities (i.e. the police) of both the origin and the destination. This is, in essence, a process to formalize, legitimize, and control migration. Transferring the hukou location from a rural area to another rural area is usually less strictly controlled and does not involve changes in hukou type. However, changing the hukou location from a rural area to an urban area necessitates also change in hukou type (i.e. *nongzhuangfei* described earlier) and is strictly controlled. Hukou location in large cities is harder to get than that in towns and small cities. As described in Chapter 2, if migration is accompanied by hukou change then the migrants are considered permanent migrants, as opposed to temporary migrants whose relocation is not accompanied by hukou change.

Although both hukou type and hukou location are part of the hukou system, the control over hukou type has declined in recent years (see "Hukou reforms"). Hukou location, however, continues to be tightly monitored and remains a key determinant of one's institutional, economic, and social statuses. The difference in status between hukou in small cities and towns and hukou in large cities persists and is substantial.

In practice, having one's hukou location in an urban area also means that the hukou type is non-agricultural. It is this kind of hukou that is most highly sought after and strictly controlled. The vast majority of peasant migrants are blocked from *nongzhuangfei* and moving their hukou to urban destinations. For the sake of simplicity, in this book I use the terms "urban hukou" and "rural hukou" to refer to the institutional statuses of, respectively, those whose hukou location is in urban areas and those whose hukou location is in rural areas.

Origins of the hukou system

As early as 685 BC (Zhou Dynasty), registration records existed that defined segments of the population by occupation (Yu 2002: 14). Between then and 1949, various forms of population registration had existed in China, mostly for the purpose of taxing and farmland allocation. Unlike these forms of registration, the hukou system was a product of central planning and was implemented under a powerful government via austere policies.

China observers and scholars have popularized two related explanations for the implementation and prominence of the hukou system under the PRC. The first explanation focuses on security and migration control (Yu 2002: 15–21). In the 1950s, re-establishing and maintaining stability in the country was deemed by the new Communist administration as an overriding priority. This goal entailed, among other means, maintaining a record of the population. Accordingly, in 1951, the Ministry of Public Security announced new regulations for managing urban residence, which stipulated that visitors to the city for more than three days must report to the Public Security Bureau (PSB) and that households, hospitals, and hotels must keep registration records. In 1955, registration of the population was extended to rural areas, via another set of regulations approved by the State Council. Thus far, however, the emphasis was on security and not on migration control. In fact, Article 90 of China's Constitution that was promulgated in 1954 states that Chinese citizens have the right to change their residence (*China Daily* 2001). Between 1949 and 1957, rural–urban migration was discouraged but not controlled.

During the first half of the 1950s, peasants migrating to cities sought to escape poverty and Soviet-type collectives in rural areas and at the same time were attracted to urban employment opportunities. By the second half of the 1950s, collectivization and crop failures had severely and adversely affected the livelihood of peasants in many parts of China, and accordingly rural–urban migration was pursued by many as a means of survival (Yu 2002: 17). Disorderly and not sanctioned by the state, these waves of rural–urban migration were labeled as "blind flows" (*mangliu*) and raised concerns over unstoppable migrants flooding cities (Cheng and Selden 1994). The Chinese state, unlike its counterparts in other developing economies that have largely failed to halt rural–urban migration, initiated a series of directives between 1953 and 1957 that culminated in the 1958 hukou regulations described earlier, and for the first time included in legal codes the different institutional statuses between rural and urban Chinese. In conjunction with other institutional mechanisms (see also "Hukou and migration control"), the hukou system severely limited rural people's ability to survive in urban areas and thus became a powerful tool for controlling migration in general and rural–urban migration in particular.

The second explanation for the implementation of the hukou system focuses less on migration control and more on the state's development strategy that favored industry over agriculture and city over countryside (Chan and Zhang 1999; Cheng and Selden 1994; Tang 1991; Tang *et al.* 1993; Wang 1997; White 1977). This strategy was in part modeled after the former Soviet Union, whose development path centered on heavy industry and on utilizing rural savings for financing industrial capital accumulation (Knight and Song 1999: 4). Arguments supporting this explanation are twofold. First, the flow of labor from rural to urban areas was blocked in order for the state to extract value from the agricultural sector to subsidize industrialization. In this view, keeping the large and cheap labor force in the countryside would ensure a supply of low-priced agricultural goods to support industrialization (*China Daily* 2001). While this argument is quite popular, it downplays the fact that rural labor surplus was very large in China and prices of agricultural goods would remain low even without control over labor flows. Second, in order to promote industrialization, the state provided urbanites with abundant support. Such support was not available to rural people. Large-scale migration to cities and rapid growth of the urban population would have bankrupted the state coffer and must therefore be prevented (Cheng and Selden 1994). Thus, peasants were expected to stay in the countryside, their agricultural production was mobilized to support urban areas and urban activities, and urbanites were designated to promote industrialization. The hukou system was one of the most powerful mechanisms of this peculiar path toward industrialization, one that has been described as "industrialization on the cheap" because it fostered "unequal exchange" between the agricultural sector and the industrial sector (Chan and Zhang 1999; Cheng and Selden 1994; Li, Si-Ming 1995; Tang *et al.* 1993; Wang 1997; see also "Hukou and migration control"). The premise of this explanation is that the state was not simply responding to *mangliu* but instead it erected barriers between the city and the countryside to advance a specific development strategy. Through the hukou system, not only did the state regulate migration but it also constructed two unequal Chinas – one urban and one rural – and divided the citizens into two unequal tiers – the privileged urban and the underprivileged rural (Cheng and Selden 1994; Christiansen 1992; Shen and Tong 1992).

Hukou and migration control

Hukou is not only a policy but also an institution and a system, because its impacts are deep and have reached every individual and aspect of Chinese society. In a narrow sense, the hukou system refers to the 1958 hukou regulations described above. In a broader sense, it is part of a host of central-planning mechanisms that affect people's lives and access to resources and benefits (Chan and Zhang 2001; *China Daily* 2001).

The central-planning mechanisms that operate in coordination with hukou regulations deal specifically with three aspects of survival in urban areas – food, employment, and welfare (Chan and Zhang 1999; Fan 2002a; Wang 1997; Yu 2002: 26–29). First, starting in 1953, the State Council used the "unified purchase and marketing" (*tonggou tongxiao*) system to allocate food (Cao 1995; Cheng and Selden 1994). Private marketing of food was prohibited. This food allocation system, which aimed at guaranteeing every urbanite sufficient food, was one of the first central-planning instruments of the PRC. It was implemented with the promulgation of various regulations throughout the 1950s, which also established that peasants must rely on their own labor for food, except in time of severe natural disaster, when the state would provide relief. When the hukou regulations came into being in the late 1950s, the food allocation system was readily available to enforce the institutional differences between urban and rural Chinese – urban hukou was an entitlement to state-allocated food, while peasants with rural hukou were on their own to generate food supply. Urban citizens were given food coupons or grain coupons (*liangpiao*) to purchase food from markets monitored by the state. In contrast, peasants did not have access to state-allocated food. Rural–urban migration without the sanction of the state was, therefore, easily curbed. In addition, under the unified purchase and marketing system, prices of agricultural goods were set low and prices of industrial goods were set high – the so-called "scissors gap" – which guaranteed that urban citizens would be supplied with low-cost food and industrialization would be subsidized with inexpensive raw materials from the countryside. In short, by controlling the prices and flows of food and agricultural goods, the state not only regulated rural–urban migration but also enabled extraction of value from the agricultural sector to the industrial sector (Knight and Song 1999: 322; Tang *et al.* 1993; Wang 1997).

Second, also in the 1950s, the state began to put in place a highly centralized and tightly controlled labor allocation system. This system, known as "unified state assignment" (*tongyi fenpei*), assigned school graduates to specific sectors, occupations, and regions according to the state's development blueprints. Likewise, workers were transferred to new jobs according to the state's plan of labor allocation. Urban work units (*danwei*)[21] were prohibited from recruiting labor from the countryside or hiring peasants. Under this system, a labor market did not exist, job mobility was low, and job search was unnecessary. On the other hand, with the "iron rice bowl," urban workers lacked incentives to improve productivity. The urban inefficiency that ensued was absorbed by the state's subsidies to cities and financed in part by the transfer of value from the countryside to urban areas, as described earlier. In urban areas, the state guaranteed employment to all in the labor force, and by doing so it was able to eliminate urban unemployment but not underemployment. In the countryside, however, peasants were on their own and the vast majority relied solely on farming for their livelihood with no access to state subsidies.

The third system related to hukou was urban welfare. Regulations on the urban welfare system promulgated in the 1950s specified many and abundant benefits available to employees of state-owned and collective enterprises. As the state sector dominated the urban economy during much of the pre-reform period, these benefits were in essence available to all urban employees and their families. They included health care, retirement benefits, and many other miscellaneous subsidies funded by the state. For example, state employees paid only nominal rent for housing. Thus, urbanites were granted entitlements protecting them in almost all aspects of life in the city. Peasants in the countryside, on the other hand, were excluded from this type of welfare provision and had to pay for all types of expenses from their farming and related activities. Given the scissors gap that suppressed prices of agricultural goods, peasants' livelihood was constantly threatened and they remained in persistent poverty.

In short, urban hukou was a guarantee to food, employment, state-subsidized welfare, and other necessities in the city. Without these necessities, survival for peasants in cities was virtually impossible. Prior to the 1980s, therefore, cities were out of reach of peasants. For decades, the hukou system, in coordination with the food-rationing, job-allocation and welfare systems, assured that peasants (a) would not be a fiscal burden to the state; (b) would be anchored to the countryside; and (c) would contribute to the transfer of value from the countryside to urban areas (Hsu 1994; Knight and Song 1999: 12; Wong and Huen 1998). Though these restrictions on rural–urban migration contradicted the 1954 Constitution that guaranteed all citizens the right to change their residence, the state justified the restrictions by defining rural–urban migration as "blind" and disorderly and by arguing that such migration would jeopardize the collective interest of the nation. For example, Luo Ruiqin, head of the Ministry of Agriculture in the late 1950s, commented in 1958: "The freedom guaranteed in the Constitution refers to the freedom under leadership rather than freedom without government. It is concerned with the freedom of the entire people rather than the absolute freedom of a minority of people" (Cited in Yu 2002: 24).

Using population registration to monitor migration is hardly a PRC invention. As described earlier, as early as the Zhou Dynasty, population registration had been used to prohibit migration (Yu 2002: 14). In the French colonies of Algeria and Vietnam, a residential registration system was used to control internal migration (Wang, Fei-Ling 2005: 157). South Africa also controlled rural–urban migration via a registration system, whose relaxation in recent years resulted in high unemployment in cities (Knight and Song 1999: 338–339).

The case of the former Soviet Union is especially noteworthy because of its socialist history and strong state, both similar to China, and because it exerted strong influence over China's development path and political organization during the early years of the PRC. The Soviet Union established a registration system in 1932 that required individuals aged 16 or above to obtain a residence permit (*propiska*), primarily to control rural–urban migration and to halt overcrowding in cities (Buckley 1995). Accordingly, migration was subject to official approval, and became another area of the society to be organized and planned, like the effect of Five-Year Plans on industrial production. Most importantly, residence registration was interwoven with government guarantees and distribution of social services, such as state-subsidized housing, education, and health care (Mitchneck and Plane 1995b). It served an important function in controlling access to social benefits and programs (Buckley 1995). Since the breakup of the former Soviet Union, most researchers' attention has shifted to the massive migration of Russians and Russian speakers from newly independent states such as Ukraine, Kazakhstan, and Uzbekistan to Russia (e.g. Pilkington 1998). Nevertheless, the legacy of Soviet-period institutions continues to be powerful. Buckley (1995: 915) well summarizes the stickiness of socialist institutions: "In any large scale socio-political transition, the institutions of the previous regime are not always compatible with the process of change and reform." Despite observations that the *propiska* system appears to be localized and ad hoc (Wang, Fei-Ling 2005: 160), it still remains in Russia, symbolizing the state's guarantee of access to official systems of distribution. The notion of a social contract between the state and individuals continues to affect migrants' calculus, so that individuals' decisions and migration processes are not only influenced by human capital and economic considerations but also by access to services and resources tied to the registration system (Mitchneck and Plane 1995a; 1995b). Similar to the case of the former Soviet Union and post-Soviet Russia, the hukou system in China is closely integrated with government-controlled distribution of resources, and it is this marriage between population registration and eligibility for state benefits and full benefits of citizenship that makes hukou an effective tool of migration control.

Criticisms of the hukou system

During the reform period and especially since the 1990s, scholars and researchers in China have engaged in heated debates over whether the hukou system should remain or be abolished and in what ways hukou reforms should take place. Changes in hukou regulations since the mid-1980s have indeed relaxed migration control (see "Hukou reforms"). Despite these changes, rural migrants in cities continue to be defined and treated as inferior citizens and are blocked from many opportunities and resources. The nature and consequences of the hukou system and the situations of rural migrants not only concern China specialists but also have caught the attention of Western media and observers (BBC News 2005; Congressional-Executive Commission on China 2005; *Los Angeles Times* 2004).

48 The hukou system

Analysts in China credit the hukou system with playing an important historical role in facilitating implementation of the central planning system, fostering orderly allocation of labor, food and health care, and controlling the size of large cities during the pre-reform period (Wan 2001; Yu 2002: 6–7; Zhang and Lin 2000). The merits of the system are, however, increasingly being challenged (Ban and Zhu 2000; Yu 2002: 6–7). Critics of the hukou system and of hukou reforms (see "Hukou reforms") are mainly concerned with issues of labor flows, rural–urban inequality, and management (Yu 2002: 377).

First, critics challenge the logic of migration control. Scholars favoring increased marketization of the Chinese economy are increasingly attracted to the neoclassical logic of free labor flows. In this view, hukou impedes the establishment and operation of a labor market, hinders efficient allocation of human resources, and in turn holds back marketization (Cao 1995; *China Daily* 2001; Yu 2002:5; Zhang and Lin 2000). When cities define eligibility for certain jobs based on hukou, they also constrain the matching between jobs and qualified people. Cai (2002) observes that hukou artificially depresses wages of rural migrants and increases wages of local labor. Similarly, Yu (2002: 5) points out that hukou undermines the market principle that links rewards to performance. He contends that restrictions over migrants' access to jobs weaken the city's ability to self-regulate and slow the pace of urban development. In addition, Yu argues that increased interactions among the population, including rural–urban interactions, are favorable for economic development, improvement of education, the labor market, and the marriage market.

Second, critics point out that the hukou system has reinforced and perpetuated inequality between the city and the countryside and between rural and urban Chinese (Hao *et al.* 1998). Numerous studies have shown that the dualistic structure embodied by the vastly different meanings of rural hukou and urban hukou has exacerbated the social and economic segregation of peasant migrants and segmented the urban labor market (Alexander and Chan 2004; Cai 2002; Cao 1995; Chan *et al.* 1999; Cheng and Selden 1994; Fan 2002a; Gu 1992; Li 1995; Solinger 1999a; Yu 2002: 40–41; Zhou 1992).[22] The bulk of peasant migrants are treated as outsiders and the prospect of their assimilation in cities is poor (Chan 1996; Fan 2002a; Solinger 1995; Zhou 1992; see also Chapter 6). Rural Chinese and urban Chinese are assigned different statuses, entitlements, treatments, and opportunities. With rare exceptions, these discrepancies are lifelong and are passed from one generation to the next, culminating in a fixated hierarchy whereby status is defined by birth and geographic origin (Yu 2002: 40–41). Critics argue not only that hukou contradicts socialist principles of equality for all but also that it is incompatible with the market principle that rewards are based on performance rather than prescribed status (Yu 2002: 5). Scholars have also compared the hukou system to barriers against immigration in Japan, Germany, and the US, and to the South African apartheid system (Alexander and Chan 2004; Roberts 1997; Solinger 1999b).[23]

The third criticism of the hukou system concerns difficulties in management. The hukou system was implemented based on the assumption of low population mobility. Under that assumption, hukou determines where one belongs and where one should be counted. As described in Chapter 2, the number and proportion of the population that are not staying in places where they are registered have increased, especially since the 1990s. This poses tremendous difficulties for the government to accurately count the population and for local authorities to properly allocate resources to serve the population. According to the 2000 census, for example, millions of people in Shanxi and Hunan were not probably registered and 130,000 people remained on Chongqing's registration record even after they were already deceased (*China Daily* 2001). Documentation of population size is extremely inconsistent; some official statistics include temporary migrants and some do not. Urban statistics that do not include temporary migrants unduly result in underestimation of urban population and level of urbanization (Wan 2001). Chan (2003), for example, points out that the actual population of Shenzhen is many times bigger than the number reported in the Urban Statistical Yearbooks because the latter do not include temporary migrants.

Despite the hukou system's drawbacks, most critics recommend reforming the system rather than abolishing it altogether (Zhang and Lin 2000; Wang 2003). The rationale for this recommendation is fourfold. First, although economic liberalization has reduced the scope of central planning, the state remains prominent in its role to guide the course of the national economy. While some strategies of the pre-reform period, such as central allocation of food, are no longer used, the state continues to employ instruments and practices inherited from that period, such as Five-Year Plans and the hukou system (e.g. Chan and Zhang 1999; Fan 2006). In this light, the logic of central planning has not been completely abandoned. Second, keeping the hukou system protects the state and local governments from bankruptcy, as they are financially incapable of extending urban benefits to all rural persons (Wang 2003). Third, urban residents often perceive migrants as competitors for jobs and state-sponsored benefits and as reasons for crime and overloaded infrastructure in cities (Messner *et al.* 2007; Jiao 2002; Wang *et al.* 2002; Zhong and Gu 2000; see also Chapter 6). Obliged to protect the interests of local residents, urban governments tend to support hukou-based policies that discriminate against migrants (Cai 2001: 331). In the same vein, policies that commodify hukou (see "Hukou reforms") encourage those who were awarded urban hukou to protect their interests and to oppose drastic hukou reforms (Wang 1997; Yu 2002: 389). Fourth, as the rural–urban gap remains large, scholars contend that some degree of migration control is necessary to prevent large cities from growing too rapidly (Cai 2002: 229; *China Daily* 2001; Wan 2001; Wang 2003). City governments are afraid that abolishing the hukou system will result in sharp pressure on employment, security, traffic and schools.

Hukou reforms

Criticisms of the hukou system and various recommendations on how to reform the system have paved the way for a variety of changes and new inventions since the 1980s. Some of the most notable changes are highlighted below.

50 *The hukou system*

- Temporary migration. In October 1984, the State Council announced that peasants working in towns would be granted the "self-supplied food grain" hukou, marking the first opening in the rigid border between city and countryside. In 1985, the Ministry of Public Security issued regulations for migrants to obtain the "temporary residence permit." In the same year, the National Congress approved the citizen's identity card as an alternative proof of identification (Yu 2002: 35). All of this, plus increased marketization of food, housing, and other daily necessities, has made it easier for rural Chinese to work and live in urban areas. Without urban hukou, however, rural migrants are considered temporary migrants even though they may have lived and worked in urban areas for an extended period of time (see also Chapter 2).
- "Selling" of hukou. A wave of local governments charging high fees[24] – ranging from several thousand yuan to tens of thousands of yuan – in exchange of hukou in small towns and cities, began in the late 1980s. Town and city governments justified this practice on the grounds that they should be compensated for extending urban benefits to migrants. Other requirements such as home purchase, investment, age, education, and skills were also used, but the specifics varied greatly from place to place (Chan and Zhang 1999; Han 1994). Yu (2002: 374) estimates that by the end of 1993 three million peasants had purchased a city or town hukou and contributed a total of 25 million yuan to the respective governments. Beginning in the mid-1990s, large cities such as Shanghai and Shenzhen started to offer "blue stamp" hukou (Wang 1997; Wong and Huen 1998).[25] Blue stamp hukou functioned like a "green card" in attracting the most desirable elements of the migrant population by providing them right of abode and certain benefits in urban areas, and in this light it is comparable to immigration policies of host countries such as Canada that favor immigrants with skills and money to invest (Cai 2001; Cao 2001; Chan and Zhang 1999). These practices commodified hukou and benefited the coffer of urban governments and the local economy (Guangdong wailai nongmingong lianhe ketizu 1995; Eastday.com 2002). In essence, urban governments are ready to make use of and exploit institutional legacies inherited from the pre-reform period – in this case, the hukou system – to the extent that such institutions are in their interest. Clearly, the vast majority of peasants were not eligible for and could not afford blue stamp hukou, which did little to mitigate the dualism fostered by the hukou system (Li 1997). Since the late 1990s, the blue stamp hukou has been increasingly replaced by other mechanisms (see below). In early 2002, Shanghai abolished the blue stamp hukou and replaced it with a new resident card designed to accommodate a selected pool of skilled workers, overseas Chinese, and foreign citizens (Eastday.com 2002). Again, the municipality is interested in granting its membership to only the elite or potential elite.
- Hukou reform in small cities and towns. In 1997, the State Council approved a pilot scheme to grant urban hukou to rural migrants who have a stable job and have resided in selected cities and towns for more than two

years (Chan and Zhang 1999; Yu 2002: 379). Unlike the blue stamp hukou and similar practices, this scheme did not require migrants to pay a large sum. The scheme was tested in 450 towns and small cities, based on which the State Council approved in 2001 plans to further expand hukou reform (*China Daily* 2001; Yu 2002: 382). Since then, the principal criteria for obtaining hukou in small cities and towns have been a fixed and legal residence and a stable source of income (Cai 2003: 210). In 1998, the State Council approved four guidelines that aimed at further relaxing urban hukou.[26] In 2003, the State Council issued a directive affirming the rights of peasant migrants to work in cities (Cai 2003: 212). Adherence to these guidelines and directives is, however, up to individual city governments.[27]

- Hukou reform in large cities. The extent and specifics of hukou reform vary greatly from one city to another. In general, the larger the city, the more difficult it is to obtain local hukou.[28] A number of large and medium-size cities such as Zhuhai, Nanjing and Xi'an have indeed relaxed their criteria for granting hukou (Cai 2002: 227).[29] Between August 2001 and June 2003, Shijiazhuang in Hebei awarded hukou to 450,000 migrants (Wang 2003). Yet, in most large cities, hukou reform is minimal; only an extremely small minority of rural migrants who satisfy stringent criteria such as educational attainment, skills, financial ability, health, and a clean criminal record are awarded local hukou and given access to urban benefits (Cai 2003: 210–211; Qiu 2001; Wang 2003; Zhang and Lin 2000; Zhong and Gu 2000). In this light, the legacy of blue stamp hukou and the logic of creaming persist. "Super-large" cities such as Beijing, Shanghai and Guangzhou, where hukou is still a primary gatekeeper, are especially resistant to hukou reform.[30] Beijing city proper hukou, for example, is required of university graduates who wish to apply for government jobs (*China Daily* 2005). Many enterprises in Beijing restrict hiring to individuals who have Beijing hukou (*Fazhi wanbao* 2006).[31] In addition, city governments can tighten the policy at their discretion. In 2002 and 2004, respectively, Guangzhou and Zhengzhou reversed their hukou reforms on the grounds that migrants overloaded the urban infrastructure (*China Daily* 2004; *Zhongguo qingnianbao* 2007). Not surprisingly, therefore, the view that hukou-based barriers are here to stay remains strong among scholars (e.g. Chan and Buckingham 2007; Wang, Fei-Ling 2005: 202).
- Hukou classification. In recent years, the distinction between agricultural hukou and non-agricultural hukou is no longer as compelling. Some provinces (e.g. Anhui, Gansu, Hunan, and Hubei) and large cities (e.g. Guangzhou and Nanjing) have, in fact, eliminated hukou classification (Congressional-Executive Commission on China 2005). Hukou location, however, continues to define one's access to resources and to a large extent one's life chances.

The new measures and guidelines described above show that the Chinese government has indeed paid greater attention to reforming the hukou system.

The Tenth Five-Year Plan (2001–2005) established the goal that by the year 2005 hukou reforms in large and medium-size cities would be completed (Cai 2002: 229). Furthermore, the Ministry of Public Security planned to replace the dualistic registration system with a unified registration system by the end of the Tenth Five-Year Plan. To date, neither of the above has been fully implemented. Still, these official endorsements indicate that the central government is increasingly concerned with tackling issues related to hukou and migration. At least two quantitative targets set in the Eleventh Five-Year Plan (2006–2010) suggest that the government would encourage migration (Editorial group 2006: 9–10; Fan 2006). First, the level of urbanization is expected to increase from about 43 percent in 2006 to 47 percent in 2010, indicating that a moderate pace of rural–urban migration is anticipated and encouraged (*Guangming ribao* 2006). Second, it is expected that by the year 2010 an additional 45 million rural workers will have shifted from rural sectors to urban sectors. Although it is too early to predict if these targets will be reached, the Eleventh Five-Year Plan has indeed legitimized a development trajectory of increased urbanization and rural–urban labor transfer, which would most certainly entail further hukou reforms.

Summary and conclusion

This chapter has focused on the hukou system as an instrument of migration control and as a source of institutional divides between the city and the countryside. In both the pre-reform and reform periods, the hukou system was used by the state toward achieving its goals for national economic development. During the pre-reform period, the state's commitment to industrialization – heavy industrialization, to be more exact – was accompanied by heavy subsidies to urban areas and transfer of value from the countryside to the city. In conjunction with institutional mechanisms that monitored the allocation of food, jobs, and welfare, the hukou system virtually wiped out rural–urban migration until the early 1980s. By binding peasants to the countryside and setting the prices of agricultural goods low, the state was successful in focusing investment and subsidies on urban areas and industrialization. Yet, the hukou system has also reinforced a dualistic system and fostered a deep divide between the city and the countryside (Cao 1995; Cheng and Selden 1994; Christiansen 1990). The system reflects the state's bias toward urbanites,[32] to whom it pledged full responsibility in terms of food, housing, work, education, and all sorts of welfare entitlements. Rural Chinese, on the other hand, were expected to be self-reliant in the countryside and were bereft of state support. The dualism that ensued is aptly described as a "crowning feature of Chinese socialism" (Wong and Huen 1998).

The hukou reforms that began in the mid-1980s have addressed two separate but related demands. First, criticisms of the hukou system necessitated a response by the government. In particular, the pre-reform tight control over migration is seen as contradictory to the goals of marketization and an increasingly globalized mode of economic development. The hukou reforms that have been implemented so far, therefore, have focused on facilitating labor flows across the country. The reforms are not designed to alleviate the inequality between rural and urban Chinese. On the contrary, many recent practices creamed off the highly educated, skilled and/or monied and absorbed the most successful and competitive peasants into the city. Extensive efforts in granting urban hukou to peasant migrants are limited to small cities and towns and these efforts are highly uneven across China. The vast majority of rural Chinese remain tied to a peasant membership and the vast majority of rural–urban migrants are denied urban citizenship. For the most part, therefore, hukou continues to be an ascribed status and defines the opportunity structure in reform-period China. Hukou reforms have not changed the fundamental barriers that block peasants, especially those with low education and poor skills, from urban membership. The small openings in the hukou wall allow only the most able and competitive to enter. Most peasant migrants can see through the glass door that separates them from the urbanites but only few can make it to the other side.[33]

A second demand that the hukou reforms have addressed is the state's developmentalist goals in the reform period. This has been elaborated in Chapter 1 and is not detailed here. In short, the hukou system constitutes an effective tool for the state to pursue a path of development that requires a large supply of cheap labor and to tap into the global market. By maintaining an institutional and social order in which peasants are inferior to urbanites, and by permitting rural–urban migration without granting urban citizenship to most migrants, the state has engendered a migrant labor regime conducive to industrial and urban development at low cost with far-reaching social consequences. These temporary migrants are, institutionally and socially, regarded as outsiders in the city. Thus, the divides that had always existed between the city and the countryside have now reproduced themselves between peasant migrants and urbanites within the city. Chapters 4 and 6 discuss in greater detail how hukou has shaped migration processes and migrants' experiences.

The story that has thus far been told focuses on the hukou system as a primary reason for the positions of peasant migrants. It illustrates the continued and prominent presence of the state, through maintaining hukou divides and above all control over urban citizenship. This hukou paradigm has, indeed, fast become a widely accepted framework for understanding migration in China. As discussed in Chapter 1, however, an approach that emphasizes hukou alone tends to downplay bottom-up, household, and individual-level perspectives. It also assumes that peasant migrants desire to stay in urban areas and want to become permanent urban residents. In Chapter 5, I shall examine this assumption as well as the role of the household. But first, in the next chapter, I shall focus on the types of migration and their connection to the two-track migration system.

4 Types and processes of migration

Introduction

Most existing perspectives in migration research assume a high degree of uniformity of migration processes. Economic and employment opportunities are widely considered the most powerful determinants of migration volume and pattern in many countries (Lewis 1954; Todaro 1976). In China, too, the role of economic considerations is prominent, which partly explains the attention paid to labor migration more than to other types of migration. Population movements are, however, highly heterogeneous. A fuller understanding of internal migration in China demands attention to a multitude of migration reasons and types, each reflecting in important ways the coexistence of state control and market mechanisms. The objective of this chapter is to examine in detail the socioeconomic characteristics and migration processes associated with different types of migration.

Migration reasons

The 1990 and 2000 censuses are valuable sources for analyzing the heterogeneity of internal migration, because they both include a question for migrants to state their primary reason for migration. Table 4.1 describes the reasons and their definitions. The options in the 1990 census are similar to those used in the 1987 One-Percent Sample Survey, while the 2000 census drops "retirement" and adds a new option, "housing change."[34] It is important to note that these "migration reasons" encompass a range of interpretations. They depict one or more of the following: motives for migration, means of migration, circumstances under which migration takes place, what migrants plan to do at the destination, and the degree of state involvement (Fan 1999).[35]

Categorization of migration reasons

There are two popular dichotomies to categorize the census migration reasons. Figure 4.1 describes the two dichotomies for the migration reasons included in the 1990 census. The economic-social (including life-cycle) dichotomy serves as an umbrella for similar dichotomies such as work-related versus non-work-

Table 4.1 Migration reasons and definitions

Reason	Definition
Job transfer	Migration due to job change, including demobilization from the military
Job assignment	Migration due to assignment of jobs by the government and recuitment of graduates from schools
Industry/business	Migration to seek work as laborers or in commerce or trade sectors
Study/training	Migration to attend schools or to enter training or apprentice programs organized by local work units
Friends/relatives	Migration to seek the support of relatives or friends
Retirement	Cadres or workers leaving work due to retirement or resignation, including retired peasants in rural areas with retirement benefits
Joining family	Family members following the job transfer of cadres and workers
Marriage	Migration to live with spouse after marriage
Housing change	Moving due to house demolition or changing house
Other	All other reasons

Sources: Population Census Office (2002: 1898); SSB (1993: 513–514, 558).

Note
The 2000 census drops "retirement" and adds "housing change."

related (Chan 1994: 115; Yang 1994: 121) and economic versus life-cycle/family (Rowland 1994) reasons. Using this approach, "job transfer," "job assignment," and "industry/business" are considered economic reasons, and "friends/relatives," "retirement," "joining family," and "marriage" are regarded as social reasons. In essence, whether work is involved in the migration process is a key factor distinguishing the two categories. However, the boundary between economic and social reasons is not always clear. "Study/training" may be considered an economic reason because of its likely correlation with future economic gains (Li 1994b; Liu and Chan 2001), but it can also be considered a life-cycle reason as it reflects moves that occur primarily at young ages (Rowland 1994; Zhai and Ma 1994). In Figure 4.1, therefore, "study/training" is given an arbitrary position between the economic and social categories. And, economic and social motivations may not be easily disentangled. "Marriage" is traditionally considered a social reason of migration and is so categorized in Figure 4.1. However, marriage may also reflect a strong economic rationale (Fan and Huang 1998; Fan and Li 2002), a notion I shall elaborate in Chapter 8.

The logic of the economic-social dichotomy is primarily based on studies of market economies where the economic motive is a key parameter describing and differentiating population movements. But the persistence of state control and socialist institutions such as the hukou system in China necessitates a second, equally important, dichotomy – that between "state-sponsored" (plan) migration and "self-initiated" (market) migration. State-sponsored migration refers to movements that are initiated by the state or those that are endorsed, recognized, and supported by the state. These are activities "within state plan" and are generally associated with hukou change so that migrants can attain residence

56 *Types and processes of migration*

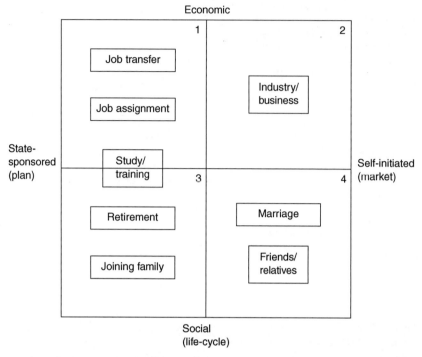

Figure 4.1 Dichotomies of migration reasons.

rights at the destination and become permanent migrants. On the other hand, self-initiated migration is primarily initiated by the migrants themselves and tends to be identified with market forces and regarded as activities "outside of state plan." Self-initiated migrants are likely not awarded local hukou in the destination and are considered temporary migrants.

Most researchers consider "job transfer," "job assignment," "joining family," "retirement," and "study/training" state-sponsored reasons, and "industry/business," "friends/relatives," and "marriage" self-initiated reasons (Liu and Chan 2001; Shen and Tong 1992: 202; Tang 1993). Both "job transfer" and "job assignment" refer to the state's allocation of human resources by, respectively, assigning school graduates (primarily university graduates) and transferring workers to particular sectors and regions (see also Chapter 3). Although the desire of migrants is indeed a factor in these moves, it is ultimately state planning, via local governments and work units, that determines the process, pattern, and feasibility of such migration. The return of demobilized military personnel, cadres, and workers from previous assignments or state-sponsored migrations is also grouped under "job transfer" (Ding 1994). "Joining family" is by definition a direct consequence of "job transfer" (see Table 4.1). Because "retirement" involves in particular the elderly retiring from the state sector and from jobs assigned by the state, it is generally considered state-sponsored migration. "Study/training" refers mainly to the pursuit of education beyond the senior secondary level. Admission to and enrollment in universities is highly competitive, is subject to state plans, and is almost always accompanied by an urban hukou. Thus, state involvement and sponsorship is prominent in "study/training" moves.

In contrast with "job transfer" and "job assignment," "industry/business" refers to self-initiated moves for working in non-state sectors, which underscores economic opportunities driven by the market (Chan 1994: 120; Li and Siu 1994; Shen and Tong 1992: 202–203). "Friends/relatives" is related to "industry/business," as migrants seek survival in the destination via their social and kinship networks. "Marriage" migration reflects primarily the desire of individuals rather than state plans. Like the economic-social dichotomy, however, the distinction between state-sponsored and self-initiated categories is not a perfect one. The extent to which "marriage" migration is endorsed and supported by the state varies. In general, "marriage" migrants who move from one rural area to another have access to resources, primarily farmland, at the destination. However, rural–urban marriage migrants, who are considerably smaller in number, have been largely excluded from enjoying state-sponsored resources in the city (see also Chapters 5 and 8).

The economic versus social and state-sponsored versus self-initiated dichotomies, as described above, are not always clear-cut. They are, nonetheless, convenient tools for interpreting the migration reasons included in the censuses.[36] A closer scrutiny of migrants' characteristics by migration reason in the next section highlights their heterogeneity and underscores the validity of the two dichotomies.

Characteristics of migrants

The statistics in Table 4.2 illustrate the migrants' demographic and socioeconomic characteristics according to the 1990 census and Table 4.3 summarizes the most prominent differences among the migration reasons.

Migrants are generally young, with a mean age of 27, and 65 percent are between 15 and 29 years old. Among the different types of migrants, "job assignment," "study/training," and "joining family" migrants are the youngest. "Job assignment" and "study/training" migrants are especially concentrated in the 15–29 age range as most are college-age students or have just finished school. A large proportion of "joining family" migrants are children (43.8 percent are under 15). "Marriage" migrants are also concentrated in the 15–29 range. "Retirement" migrants are the oldest.

Sex ratios show that as a whole there are more male migrants than female migrants. This holds true for most migration types, except "friends/relatives," "joining family," and "marriage," which have a female majority. This appears to support the conventional view that female migration is largely social in nature and that most female migrants are tied movers. In Chapter 5, I shall examine evidence that challenges this view.

Marital status is highly related to the age distribution of migrants. A very small proportion of "study/training" migrants and the vast majority of "retirement" migrants are married. Not surprisingly, almost all "marriage" migrants are married.

Table 4.2 Characteristics of migrants by migration reason, 1990 census

	Job transfer	Job assignment	Industry/ business	Study/ training	Friends/ relatives	Retirement	Joining family	Marriage	All migrants
Number ('000)	4,246	2,386	8,326	4,543	3,442	570	3,882	4,874	35,350
%	12.0	6.8	23.6	12.9	9.7	1.6	11.0	13.8	100.0
Age									
Mean (years)	33.7	24.4	27.7	20.6	31.1	58.4	20.6	26.1	27.0
15–29 (%)	41.7	95.0	70.1	98.3	40.1	0.5	35.1	85.1	65.0
Sex ratio	236.9	261.4	247.2	181.7	76.5	468.1	66.2	10.7	127.4
Marital status: married (15+) (%)	79.1	36.0	47.6	2.6	47.6	91.0	46.3	99.8	53.0
Education: senior secondary and above (6+) (%)	54.1	75.7	11.0	97.8	10.3	13.6	14.3	8.3	32.0
Origin (%)									
Streets	47.1	54.1	4.4	22.6	10.0	54.3	21.3	3.4	19.6
Towns	29.5	18.5	10.3	31.6	16.8	21.8	24.5	10.8	18.9
Townships	23.4	27.4	85.3	45.8	73.2	23.9	54.2	85.8	61.5
Destination (%)									
Cities	57.6	62.3	61.4	82.0	61.7	34.9	56.2	31.2	56.6
Counties	42.4	37.7	38.6	18.0	38.3	65.1	43.8	68.8	43.4
Hukou type: agricultural (%)	9.1	14.9	92.5	4.1	64.8	10.2	35.2	91.4	50.1
Hukou location: permanent migrants (%)	81.7	81.7	4.0	90.2	36.8	72.0	67.6	61.4	54.1
Interprovincial (%)	38.9	26.5	40.6	20.7	37.1	32.4	32.8	31.3	32.6

Sources: 1990 census one percent sample.

Table 4.3 Summary comparison among migration reasons

	Job transfer	Job assignment	Industry/business	Study/training	Friends/relatives	Retirement	Joining family	Marriage	All migrants
Demographic/human capital									
Age	older	very young	young	very young	older	old	young	young to very young	young
Gender	male majority	male majority	male majority	male majority	female majority	predominantly male	female majority	predominantly female	male majority
Educational attainment	average	high	very low	very high	very low	very low	very low	very low	low
Spatial									
Rural vs urban	urban–urban	urban–urban	rural–urban	urban–urban and rural–urban	rural–urban	urban–rural	all combinations	rural–rural	rural–urban
Intraprovincial vs interprovincial	majority intraprovincial	majority intraprovincial	more intraprovincial than average	more intraprovincial than average	majority intraprovincial	majority intraprovincial	majority intraprovincial	majority intraprovincial	majority intraprovincial
Hukou status									
Permanent vs temporary	permanent	permanent	temporary	permanent	mostly permanent	permanent	mostly permanent	mostly permanent	mixed
Categorization									
Economic vs social	economic	economic	economic	economic/social	social	social	social	social	mixed
State-sponsored vs self-initiated	state	state	self	state	mostly self	state	mostly state	mixed	mixed

With 32 percent at senior secondary level or above, migrants as a whole have higher educational attainments than those of the general population (see also Table 4.4). The educational attainments of "job assignment" and "study/training" migrants are especially high, and "job transfer" migrants are more highly educated than migrants as a whole. All other types of migrants have educational attainments significantly below the average.

Migrants primarily move from rural origins to urban destinations (see also Chapter 2 and Table 2.3). "Industry/business" and "friends/relatives" migrants, in particular, are overwhelmingly from rural origins and to urban destinations. Although the vast majority of "marriage" migrants have rural origins, most move to rural rather than urban destinations. In contrast, "retirement" migrants are mostly urban–rural migrants. "Job transfer" and "job assignment" migrants are largely urban–urban migrants. "Study/training" migrants may come from rural or urban areas but their destinations are overwhelmingly urban areas where advanced education institutes concentrate.

As expected, interprovincial moves, which entail longer distance and more demanding intervening obstacles than intraprovincial moves, account for a minority of migrants (32.6 percent). But "industry/business" migrants are more likely

Table 4.4 Comparison among non-migrants, permanent migrants and temporary migrants, 1990 census

	Non-migrants	Permanent migrants	Temporary migrants	All migrants
Age				
Mean (years)	31.9	26.3	27.9	27.0
15–29 (%)	34.0	68.4	61.1	65.0
Sex ratio	105.1	123.2	132.5	127.4
Marital status: married (15+) (%)	68.7	50.6	55.7	53.0
Male	66.4	38.6	51.2	44.4
Female	69.3	65.6	61.9	63.9
Education: senior secondary and above (6+) (%)	9.9	47.2	13.9	32.0
Origin (%)				
Streets	–	28.2	9.4	19.6
Towns	–	23.7	13.3	18.9
Townships	–	48.1	77.3	61.5
Destination (%)				
Cities	23.8	55.6	57.7	56.6
Counties	76.2	44.4	42.3	43.4
Hukou type: agricultural (%)	80.4	23.5	81.4	50.1
Interprovincial migrants (%)	–	28.1	38.0	32.6
Sources				
Eastern	–	35.5	44.7	38.8
Central	–	37.6	31.1	34.3
Western	–	26.9	24.2	26.9

Source: 1990 census one percent sample.

(40.6 percent) than other migrants to undertake interprovincial moves, reflecting economic opportunities in more prosperous provinces that target migrants from poorer provinces (e.g. Fan 1996). "Study/training" migrants, in contrast, have the least likelihood to undertake interprovincial moves (20.7 percent), showing that distance remains a profound obstacle for moves to pursue education.

Hukou type and location depict a strong correlation between the state-sponsored versus self-initiated dichotomy on one hand and hukou status on the other. State-sponsored migrants – "job transfer," "job assignment," "study/training," "retirement," and "joining family" migrants – are mostly given hukou at the destination and become permanent migrants. Their hukou type is more likely non-agricultural than agricultural. Most "industry/business" and "friends/relatives" migrants, in contrast, are self-initiated, do not have hukou at the destination and remain temporary migrants. "Marriage" migrants are unique, because they are, strictly speaking, self-initiated migrants and yet the vast majority of them are permanent migrants. This reflects marriage migrants' rural destinations, where the granting of local hukou to in-migrants is less strictly controlled than that in urban areas.

In summary, other than "marriage" and "retirement" migrants, permanent migrants are mostly correlated with higher educational attainments, urban origins, and urban destinations, and temporary migrants are associated with low educational attainments, rural origins, and urban destinations. These relationships suggest that the state is selectively awarding urban and skilled migrants permanent residence at the destination while relegating rural and less-qualified migrants to unofficial and temporary statuses. They reinforce the validity of the state-sponsored versus self-initiated dichotomy, in addition to the conventional economic-social categorization, for understanding migration types in China. The hukou system has played an important stratification role by engineering a two-track migration system, whereby a superior track is set aside for urbanites and the elite and an inferior track is designated for peasants and the less advantaged.

1990 and 2000 census comparisons

According to the 1990 census, "industry/business" accounts for the largest proportion of intercounty migrants, followed by "marriage," "study/training," "job transfer," "joining family," "friends/relatives," "other," " job assignment," and "retirement" (Figure 4.2).[37] By the 2000 census, the relative proportions of migration reasons have changed considerably. The share by "industry/business" has increased from 23.6 percent to 46.4 percent and as a result it has become the dominant reason for migration. "Study/training" is a distant second, followed by "joining family," "marriage," "friends/relatives," "housing change," "other," "job transfer," and "job assignment." The increased prominence of "industry/business," combined with the declining proportions of "job transfer" and "job assignment" migrations, shows that market mechanisms have become the overwhelming means, compared to state-controlled channels, for job-related moves (Ding *et al.* 2005; Wang 2004). Between the two censuses,

62 *Types and processes of migration*

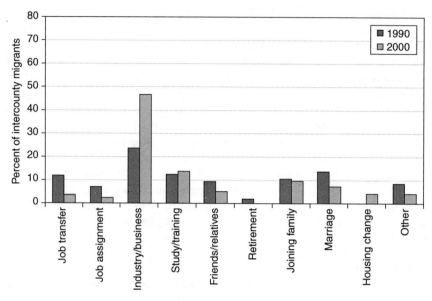

Figure 4.2 Migration reasons, 1990 and 2000 censuses (sources: 1990 census one percent sample; Liang and Ma (2004)).

the proportion attributed to state-sponsored reasons has declined from 44.2 percent to 29.4 percent, while that attributed to self-initiated reasons has increased from 47.1 percent to 58.7 percent. Economic reasons (excluding "study/training"), whose proportion has increased from 47.1 percent to 52.1 percent, have become more prominent than social reasons, whose share has declined from 32.1 percent to 22.3 percent (Wang 2004; Yang 2004).

Figures 4.3 and 4.4 describe the contributions of different migration reasons to permanent migrants and temporary migrants and how they have changed between the two censuses. Overall, the relationships between migration reasons and hukou status are consistent with the descriptions in Table 4.3. Among permanent migrants, "study/training," "job transfer," and "marriage" are the three leading reasons according to the 1990 census (Figure 4.3). In the 2000 census, the relative proportions of "study/training" and "marriage" migrants have both increased, while the proportions of most other reasons have declined. The decline is most notable for "job transfer." "Housing change," despite being a new migration reason in the 2000 census, accounts for more than 10 percent and a greater proportion of permanent migrants than "job transfer," "job assignment," and "joining family." These changes suggest that the traditional means by which migrants obtain hukou at the destination – state-allocated employment – are increasingly giving way to processes that emphasize individual efforts. "Study/training," which accounts for almost 40 percent of permanent migrants in the 2000 census, highlights in particular performance and skills.

Among temporary migrants (Figure 4.4), the dominance of "industry/business"

Types and processes of migration 63

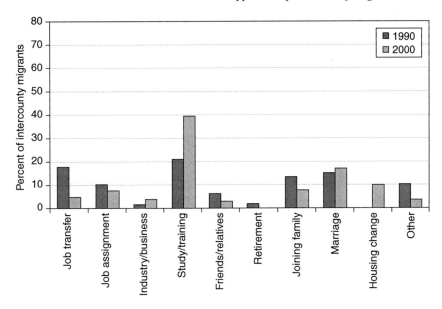

Figure 4.3 Permanent migrants by migration reason, 1990 and 2000 censuses (sources: 1990 census one percent sample; Liang and Ma (2004)).

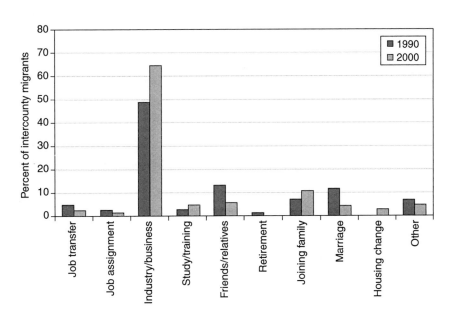

Figure 4.4 Temporary migrants by migration reason, 1990 and 2000 censuses (sources: 1990 census one percent sample; Liang and Ma (2004)).

64 *Types and processes of migration*

migrants has increased further, from 49.3 percent in the 1990 census to 65 percent in the 2000 census. Most other reasons account for less than 10 percent of temporary migrants. The proportion of "joining family" migrants has increased slightly, from 7.8 percent to 10.7 percent, which supports the observation that more temporary migrants are bringing their families to cities (Zhou 2004; see also Chapter 9).

Permanent migrants and temporary migrants

1990 census

A comparison between permanent migrants and temporary migrants can illustrate further the two-track migration system described earlier. Table 4.4 summarizes the differences among non-migrants, permanent migrants, and temporary migrants according to the 1990 census. As expected, migrants are younger than non-migrants. Migrants have higher sex ratios than non-migrants, showing that men have higher mobility than women. Migrants are less likely than non-migrants to be married, reflecting the former's younger age. A considerably smaller proportion of male permanent migrants (38.4 percent) compared to female migrants (65.6 percent) are married, probably due to the former's higher representation among young, "job assignment" and "study/training" migrants.

The differentials in educational attainment among the three groups are striking. Migrants are more highly educated than non-migrants, but permanent migrants are significantly more highly educated than temporary migrants. Respectively, 47.2 percent of permanent migrants and only 13.9 percent of temporary migrants have had education at senior secondary or above levels. The three groups' hukou types are even more divergent – 80.4 percent of non-migrants, 23.5 percent of permanent migrants and 81.4 percent of temporary migrants have agricultural hukou. Non-migrants' high proportion is attributable to their rural locations, as 76.2 percent of them live in counties. As the majority of temporary migrants live in cities (57.7 percent), their overwhelming proportion having non-agricultural hukou reflects not so much where they live but their rural origin (77.3 percent from townships) and their inferior and temporary status in the eyes of the state.

A higher proportion of temporary migrants (38 percent) than permanent migrants (28.1 percent) undertake interprovincial moves, which is probably related to the former's higher representation among "industry/business" moves, which tend to be of longer distances. Among all interprovincial migrants, higher proportions of permanent migrants than temporary migrants originate from the central and western regions, which may reflect "study/training" migrants moving to large coastal cities, return migrants ("job transfer") back to the eastern region, and "marriage" migrants undertaking mainly west-to-east moves (see also Chapter 8).

The above suggests that substantial contrasts exist among the three groups and especially between permanent migrants and temporary migrants. Table 4.5 summarizes results of a logistic regression, which aims at assessing the most prominent differences between permanent migrants and temporary migrants. Statistical analyses such as regression permit assessment of the independent effects of determinants. For example, in order to evaluate the strength of institutional factors it is important to isolate the contribution of human capital factors.

The dependent variable is coded 1 for permanent migrants and 0 for temporary migrants aged 15 or older. In essence, the analysis evaluates how well the independent variables predict migration that is accompanied by hukou change

Table 4.5 Logistic regression on permanent and temporary migrants, 1990 census (15+)

Independent variable	Standardized regression coefficient	Wald statistic	Odds ratio
Demographic			
AGE	−0.43	2,551.29****	0.98
* GENDER (male = 1)	0.24	662.22****	1.29
* MARRIED	0.11	120.94****	1.13
EDUCATION (reference: below senior high)			
*SENIOR SECONDARY	0.31	1,259.89****	1.66
*ABOVE SENIOR SECONDARY	1.20	10,588.59****	4.86
Location			
SOURCE (reference: central and western regions)			
*INTRAPROVINCIAL	0.33	1,056.54****	1.46
*INTERPROVINCIAL	−0.02	5.20**	0.96
(eastern region = 1)			
* URBAN	0.51	2,955.53****	1.75
(urban origin type = 1)			
Access			
*SELF-INITIATED	−3.06	39,711.30****	0.02
("industry/business" = 1)			
Model chi-square		163,879.25****	
−2 log likelihood with intercept		447,271.78	
−2 log likelihood of model		283,392.54	
ρ^2		0.37	
Percent correctly classified		78.91	
Number of cases		324,458	
Degree of freedom		9	

Source: 1990 census one percent sample.

Notes
Dependent variable: 0: temporary migrants; 1: permanent migrants.
* dummy variable
Significance levels: **: 0.05; ***: 0.01; ****: 0.001.
$\rho^2 = 1 - (-2$ log likelihood of model$/-2$ log likelihood with intercept).

versus migration not accompanied by hukou change. Three groups of independent variables are included – demographic, location, and access. It is expected that migrants who are more highly educated, from urban locations, and have greater access to state-sponsored opportunities are more likely permanent migrants, while individuals who are less highly educated, from rural areas, and undertake self-initiated moves are more likely temporary migrants.

A high percentage of correctly classified observations (78.9 percent) and a reasonable ρ^2 (0.37) suggest that the independent variables as a whole are quite successful in identifying the salient differences between permanent migrants and temporary migrants. Because odds ratios are unit-dependent, they are reported for reference purposes only. Instead, the size of standardized regression coefficients and associated significant tests are more reliable indicators of the relative importance of the independent variables in predicting the permanent migrant versus temporary migrant outcome.

The overall results support the expectations described earlier. All independent variables have significant coefficients with expected signs. The most influential independent variable is SELF-INITIATED (dummy variable coded 1 for "industry/business" migration and 0 for other types of migration), whose coefficient is the largest (in absolute value) among all independent variables. Its negative sign indicates that self-initiated migrants are more likely temporary migrants. The second-most-influential independent variable is ABOVE SENIOR SECONDARY. Another education variable SENIOR SECONDARY is also significant and positively related to permanent migration but its coefficient is much smaller than that of ABOVE SENIOR SECONDARY. This suggests that university-level education is especially influential in increasing the likelihood of permanent migration (versus temporary migration). URBAN, coded 1 for migrants whose origin is a street or town (versus township), is the third-most-influential independent variable. Its positive coefficient indicates that migrants from urban origins are more likely permanent migrants.

In addition, the analysis shows that younger persons (AGE), men (GENDER), married persons (MARRIED), intraprovincial migrants (INTRAPROVINCIAL), and those from the central and western regions (INTERPROVINCIAL) are more likely permanent migrants than temporary migrants, confirming the observations made earlier via descriptive statistics.

These results emphasize two distinct roles of the state. First, the state stratifies the migrants by awarding permanent residence rights to the most selective and competitive – the group that is young, mostly male, highly educated, and from urban origins. Second, the role of state sponsorship is very important in determining whether a migrant obtains permanent residence rights, *independent* of his/her attributes such as age, gender, educational attainment, and rural/urban origin. Permanent migrants' ability to obtain local hukou is not only a result of their high education and urban background; likewise, temporary migrants' low education and rural backgrounds are not the only reasons for their not obtaining local hukou. This observation underscores structural and institutional effects whereby hukou status is not simply a reward to the skilled and competitive but is above all a gate-keeping mechanism which reinforces existing inequalities due to personal attributes and which further stratifies the migration system into two tracks – one for state-sponsored privileged migrants and another for self-initiated migrants.

1990 and 2000 census comparisons

Table 4.6 summarizes the characteristics of permanent and temporary interprovincial migrants. The contrasts between permanent and temporary migrants observed earlier for intercounty migrants (Table 4.4) similarly describe interprovincial migrants. Permanent interprovincial migrants, compared to temporary interprovincial migrants, are more highly educated, less likely to have agricultural hukou, and more likely to be from urban origins.

Interprovincial migrants, however, have a much higher sex ratio than intercounty migrants, the former being 142.1 and the latter 127.4, according to the 1990 census. The gap is especially large for temporary migrants; the sex ratio is 155.8 for interprovincial migrants and 132.5 for intercounty migrants. Since interprovincial migration generally involves longer distances than intercounty migration, this supports the conventional view that men are more likely than women to engage in long-distance migration. A notable change between the 1990 and 2000 censuses is the sharp decline of the sex ratio of interprovincial migrants from 142.1 to 109.8, reflecting a considerable increase in mobility among women – more so than men (see Chapter 5). This change describes both permanent and temporary migrants, whose sex ratios have declined from respectively 128.3 to 105.6 and 155.8 to 110.5.

While the average age of interprovincial migrants has not had significant change, the 2000 census reports a larger age gap – almost two years – between permanent migrants and temporary migrants. The younger age of permanent migrants likely reflects the increased prominence of "study/training" migrants among them, as described earlier and in Figure 4.3. Related to this is a decline of the proportion married among permanent migrants from 55.5 percent to 47.8 percent between the two censuses.

Data for educational attainment depict an apparently paradoxical trend. Although the proportions of permanent and temporary interprovincial migrants with senior secondary and above level of education have both increased, from respectively 46.5 percent to 55.9 percent and 12.7 percent to 17.4 percent, the proportion of the two groups combined for this indicator has actually declined, from 28.5 percent to 22.5 percent, between the 1990 and 2000 censuses. This is partly because the volume of temporary migrants, who have much lower educational attainment than permanent migrants, has increased much faster than that of permanent migrants (see Chapter 2 and Table 2.2), thus depressing the average educational attainment. Moreover, the gap in educational attainment between the two groups has enlarged. Between the two censuses, permanent interprovincial migrants gained 9.4 percent points while temporary interprovincial migrants gained only 4.7 percent points.

Not only have migrants as a whole had a decline in educational attainment,

Table 4.6 Comparison between permanent and temporary interprovincial migrants, 1990 and 2000 censuses

	1990 census			2000 census		
	Permanent migrants	Temporary migrants	All migrants	Permanent migrants	Temporary migrants	All migrants
Number (million)	5.4	6.2	11.5	4.2	27.5	31.7
% of all migrants	46.6	53.4	–	13.2	86.8	–
Age						
Mean (years)	27.2	27.6	27.5	25.2	27.2	26.9
15–29 (%)	62.8	63.7	63.3	71.5	62.7	63.8
Sex ratio	128.3	155.8	142.1	105.6	110.5	109.8
Marital status: married (15+) (%)	55.5	55.3	55.4	47.8	54.1	53.2
Male	45.4	48.5	47.1	30.8	55.6	52.4
Female	68.7	66.2	67.5	65.6	52.4	54.1
Education: senior secondary or above (6+) (%)	46.5	12.7	28.5	55.9	17.4	22.5
Origin (%)						
Streets	42.7	12.0	26.4	36.0	9.3	12.8
Towns	17.6	10.5	13.8	37.7	49.5	48.0
(Residents' committees)	–	–	–	13.8	6.2	7.2
(Villagers' committees)	–	–	–	23.9	43.3	40.8
Townships	39.6	77.5	59.8	26.3	41.2	39.2
Destination (%)						
Cities	50.2	51.3	50.8	57.6	52.7	53.4
Towns	–	–	–	9.3	22.1	20.4
Counties	49.8	48.7	49.2	33.0	25.2	26.2
Hukou type: agricultural (%)	26.2	80.6	55.1	34.7	85.6	78.9
% of intercounty migrants	28.1	38.0	32.6	20.7	46.7	40.1

Sources: 1990 census one percent sample; 2000 census 0.1 percent interprovincial migrant sample.

their proportion with agricultural hukou has also increased, from 55.1 percent to 78.9 percent. While part of that may be due to the increased proportion of temporary migrants, the vast majority of whom have agricultural hukou, the proportion with agricultural hukou has in fact increased for both permanent and temporary migrants, from respectively 26.2 percent to 34.7 percent and 80.6 percent to 85.6 percent, between the two censuses. Origin data further reinforce the observation that the proportion of migrants from rural areas has increased. The two censuses' origin and destination data are difficult to compare (see Chapter 2), but the proportions of permanent and temporary migrants originating from streets have both declined. And, if one considers both townships and villagers' committees of towns as rural places, then respectively 50.2 percent and 84.6 percent of permanent and temporary interprovincial migrants in the 2000 census are from rural origins. Both permanent and temporary migrants continue to favor urban destinations over rural destinations.

Data on educational attainment, hukou type, and origin and destination types reviewed above point to two major findings. First, the two-track migration system described earlier for the 1990 census has continued unabated, as reflected in the 2000 census. If anything, the gap in socioeconomic and human capital indicators between permanent and temporary migrants has further widened over time. Second, increasing proportions of migrants, especially temporary migrants, have rural origins and agricultural hukou. This shows that not only have peasant migrants increased their shares of and impacts on internal migration in China but they continue to hold rural hukou instead of urban hukou. These two findings have implications for both the processes of migration and experiences of migrants. They support the notion that, as migration networks develop and mature, the selectivity of migrants declines, that is, individuals who might not in the past have decided or managed to migrate have joined the migration streams (Liang 2001b; Massey *et al.* 1993). The findings also reinforce the argument made earlier that peasant, temporary migrants are at the bottom of the socioeconomic hierarchy at the destination, which reflects both their disadvantaged human capital as well as institutional barriers in urban areas, and suggest that their inferior positions have persisted, if not deteriorated, over time.

Considerations and processes of migration

Analysis of migration reasons above highlights the different types of migration and sheds light on the motives and means of migration. Specialized surveys and qualitative information can yield further insights on the considerations and processes of migration. To this effect, below I draw from the 1998 Guangzhou survey and the 1995 Sichuan and Anhui Interview Records (see Chapter 1).

1998 Guangzhou survey

Table 4.7 summarizes the results of three questions put to permanent and temporary migrants in the Guangzhou survey. Among motives for migration, "job

Table 4.7 Migration considerations, 1998 Guangzhou survey

	Permanent migrants	Temporary migrants
Motive for migration (%)		
Job search	38.5	55.7
Study	28.8	1.1
Family/marriage	15.1	3.2
Increase income	11.7	37.7
Other	6.0	2.2
Reason for leaving origin (%)		
Study	39.3	0.6
Low income	28.0	63.5
Family	16.0	8.8
Few jobs	5.7	19.1
Other	11.0	8.1
Reason for choosing Guangzhou* (%)		
Higher wages	30.8	23.2
Ease in finding jobs	19.0	32.1
Family/relatives	17.7	26.2
Proximity to origin	12.9	13.0
Other	19.6	5.5

Sources: 1998 Guangzhou survey; adapted from Fan (2002a) with permission.

Note
* Multiple responses are permitted. All responses are included in percentage calculation.

search" is the leading answer for both permanent and temporary migrants, supporting the census finding that economic reasons are prominent in migration decision-making. Economic gain is an especially important consideration for temporary migrants, of whom a total of 93.4 percent choose "job search" or "increase income" as their motive. Push-pull factors further reinforce this point. A total of 82.6 percent of temporary migrants cite "low income" or "few jobs" as the primary reason for leaving the origin. "Higher wages" and "ease in finding jobs" account for a total of 55.3 percent of the responses by temporary migrants as the reason for choosing Guangzhou as the destination.

Among permanent migrants, economic motives are important but are not as overwhelming as in the case of temporary migrants. "Study" is the second-most-important motive for migration and the most important reason for leaving the origin for permanent migrants. Among reasons for migration, "increase income" ranks fourth, after "family/marriage;" and among reasons for leaving the origin, "few jobs" ranks last and is less important than "family." Nonetheless, "higher wages" and "ease in finding jobs" are indeed leading reasons for permanent migrants to choose Guangzhou as the destination.

The above comparison supports the notion that decision-making of temporary migrants is heavily driven by monetary return through employment, while that of permanent migrants is more evenly distributed between job-related economic reasons and social or lifecycle considerations. In addition, the strong economic push from the origin for temporary migrants, combined with observations that they are more likely than permanent migrants to originate from rural and less-developed areas (see "Permanent migrants and temporary migrants"), suggests that they are overwhelmingly "upward" movers responding to the large gaps in income and employment opportunities between their origins and Guangzhou (Ma et al. 1997). Permanent migrants, in comparison, are represented by both "upward" and "lateral" movers, including those from urban and more-developed origins. For both permanent and temporary migrants, however, "proximity to origin" is not among the most important reasons for choosing Guangzhou as the destination, suggesting that the economic pull of Guangzhou is powerful enough to overcome friction of distance.[38]

Among reasons for choosing Guangzhou as the destination, "family/relatives" ranks behind "higher wages" and "ease in finding jobs" but does account for significant proportions of the responses. Among temporary migrants, "family/relatives" accounts for 26.2 percent of the responses. This result highlights the role of social networks in the processes of migration; the availability of family, relatives and friends as sources of information and assistance is an important factor of migration decision-making, especially for temporary migrants.

1995 Sichuan and Anhui Interview Records

In the 1995 Sichuan and Anhui Interview Records, "increase income" and "insufficient farmland for food" are respectively the most and second-most frequently cited reasons for migration for both male and female migrants (Table 4.8). In addition, "surplus labor" ranks second highest for female migrants, indicating that their labor is not needed in farming because there is already plenty of rural labor relative to available farmland. This 26-year-old Sichuan woman's remark is consistent with many migrants' responses: "In our family there is plenty of labor but very little land. I had nothing to do at home, so I decided to find migrant work" (NNJYZ 1995: 121–122). These results reinforce the observation that seeking employment for monetary gain is the most important objective of peasant migrant workers and they pursue migrant work in response to the wide economic gap between the origin and destination. It is notable that "broaden horizon and learn skills" ranks quite highly among reasons for migration, suggesting that migration is also used as a means toward self-improvement. This supports observations in other studies that highlight peasant migrants' – especially young, female migrants' – desire to increase exposure (e.g. Jacka 2006: 134–138). Direct economic betterment through migrant work, however, remains the overriding motive.

There are similarities and differences between the 1995 Sichuan and Anhui Interview Records and the 1998 Guangzhou survey in the reasons for choosing the migration destination. In both cases, proximity to origin is not an important factor. Respondents in the Sichuan and Anhui Interview Records migrate long distances to work (Fan 2004b). In contrast to the Guangzhou survey, income and job factors – "better income" and "job opportunities" – are not the most prominent reasons for choosing the destination. Instead, the majority of the

Table 4.8 Migration considerations and information, 1995 Sichuan and Anhui Interview Records

	Men	Women
*Reason for migration**		
Insufficient farmland or food	41.9	31.8
Education fees for family members	17.9	18.2
To pay off debts	9.8	11.4
Surplus labor	6.0	2.3
Increase income	6.0	18.2
Build new house	6.0	0.0
Marriage expenses	4.3	9.1
Broaden horizon and learn skills	3.4	2.3
Dislike farmwork	3.0	2.3
Other	1.7	4.5
Reason for choosing destination		
Fellow villagers, relatives or friends	58.4	60.7
Better income	14.6	10.7
Job opportunities	10.9	17.9
Proximity to home village	3.6	3.6
Other	12.4	7.1
Migration company		
Family or relatives	42.2	34.6
No company	30.2	7.7
Fellow villagers	26.7	57.7
Other	0.9	0.0
Information about migrant work		
Self	33.6	17.9
Family or relatives	27.9	28.6
Fellow villagers	23.6	25.0
Recruitment	3.6	10.7
Other	9.3	0.0

Sources: 1995 Sichuan and Anhui Interview Records.

Notes
* Multiple responses are permitted. All responses are included in percentage calculation.

migrants in the Interview Records select "fellow villages, relatives or friends" as their primary reason. This underscores the notion that rural, temporary migrants rely heavily on social networks as a source of information and support, which reflects also their lack of more formal affiliations with and institutional support from the state (Solinger 1999a: 242).

Migration company and information about migrant work reinforce further the important role of social networks for rural migrants (Table 4.8). Respectively 68.9 percent and 92.3 percent of male and female migrants undertake their migration together with fellow villagers, family members or relatives. While 30.2 percent of male migrants travel alone, only 7.7 percent of female migrants do so, suggesting that company is crucial to women's mobility, and the journey from home is an impeding factor. Many rural Chinese, especially women, have never traveled to places far from their home village, and must rely on people they know who are familiar with the route. Furthermore, there is a widespread and valid perception that the journey to and from the destination is dangerous. Migrants are especially wary of robbers. Traveling with someone they trust is a risk-reducing strategy. To this 32-year-old Sichuan woman, the trip from her home village to Wujiang, Jiangsu, where she works as a machinery factory worker, was almost overwhelming:

> I left the village right after the Spring Festival. I went with my cousin and five other fellow villagers. We took the train to Zhengzhou, changed to Xuzhou, Nanjing, and then Suzhou. From Suzhou we took the bus to this town in Wujiang. That was my first time journeying a long way from home. We made so many connections that I felt dizzy. The entire trip I followed my cousin closely and didn't dare to move around lest I got lost. The trains and buses were very crowded and disorderly. We had some problems with the connections, but luckily we didn't come across any robbers during the trip.
>
> NNJYZ 1995: 2–4

This Anhui woman's encounter with robbers has likely added to her concern over the migration journey:

> The journeys to work and back home are unsafe. My fellow villager and I [traveled from Changzhou to Nanjing and] took the bus to return from Nanjing to the village. At 3 in the morning, four robbers got onto the bus and searched our belongings. Fortunately, my bag was with the driver and wasn't taken away. But my fellow villager's valuables were robbed. Not only that, the robbers beat her up because she argued with them.
>
> NNJYZ 1995: 426–428

Not only do fellow villagers, family members, and relatives provide company during the migration journey, they are also the main source of job information for the majority of migrants (51.5 percent of male migrants and 53.6 percent of female migrants) (Table 4.8). While 33.6 percent of male migrants find jobs on their own, only 17.9 percent of female migrants do so. Recruitment – referring to formal recruitment by employers and employment agencies of labor in rural areas – accounts for the employment of 10.7 percent of female migrants but only 3.6 percent of male migrants. These results suggest that female migrants are more reliant than male migrants on information provided by others, and also that recruiters may focus their efforts on female migrants more than male migrants. All in all, the findings support the notion that, other than the more experienced migrants, most rural Chinese have few means to obtain and evaluate information about the urban labor market. In Chapter 6, I shall examine further the labor market experiences of peasant migrants.

Summary and conclusion

In this chapter, I have shown that in addition to the conventional economic versus social (life-style) dichotomy, the state-sponsored (plan) versus self-initiated (market) dichotomy is fundamental for conceptualizing population movements in China. The latter is strongly correlated with the hukou system. Permanent migrants are primarily tied to institutional opportunities via government channels or admission to education institutes, while temporary migrants are largely associated with market or "outside of state plan" processes of migration. The state plays a gate-keeping role in determining access to hukou at migrants' destination, independent of individual attributes such as age, gender, education, and origins. In other words, through institutional means, the state has reinforced stratification among migrants by rewarding state-sponsored migrants, who are more highly educated than self-initiated migrants, hukou and related benefits at the destination.

Since the 1980s, economic reasons for migration have gained importance relative to social reasons, and self-initiated reasons have gained importance relative to state-sponsored reasons. The roles of economic motives and market-driven processes, especially in the form of seeking industrial and services work, are increasingly defining the patterns and processes of migration in China. Nonetheless, the two-track migration system – comprising permanent migrants and temporary migrants – continues. Over time, the socioeconomic gap between permanent and temporary migrants has widened. This reflects the decline in selectivity among temporary migrants as more rural Chinese join migration streams, as well as the persistent inferiority of human capital and institutional positions of temporary migrants relative to permanent migrants.

Considerations and processes of migration highlight the roles of economic motives and social networks among temporary migrants more than permanent migrants. The overriding objective to increase income has important implications for peasant migrants' strategies, which is the subject of Chapter 5. Social networks are prominent not only in peasants' migration process but also in their job search and urban experiences, which will be elaborated in Chapter 6.

5 Gender and household strategies

Introduction

The prevailing assumption about gender differentials in migration is that men play more central roles than women in population movements. The widely held notions that men are more mobile than women, that men travel longer distances than women, and that men move primarily for economic reasons and women for social reasons explain researchers' greater attention on the mobility of men than on that of women (e.g. Gu and Jian 1994: 24; Liu 1990; Wang and Hu 1996: 91–92; Xu and Ye 1992; Yang 1991). Some of these notions are popular in Western literature and can be traced back to E.G. Ravenstein's (1885: 197) observation that "females are more migratory than males within the kingdom of their birth, but ... males more frequently venture beyond." In China, these notions are reinforced by the persistence of sociocultural traditions that downplay women's roles in society (e.g. Li, Shuzhuo 1993; Yang 1994: 201; Yu and Day 1994; Zhang 1995; see Chapter 1). Thus, women's mobility has remained peripheral to the mainstream scholarly and policy discourses on internal migration in China.

Since the mid-1990s, more studies on the role of gender in migration in China have appeared (e.g. Cai 1997; Davin 1997; 1998; 1999; Entwisle and Henderson 2000; Fan 2000; 2004a; 2004b; 2004c; Gaetano and Jacka 2004; He and Gober 2003; Huang 1999; Jacka 2006; Lee 1998; Wang 2000; Xu 2000; Yang and Guo 1999; Zhang 2001; Zhang et al. 2005). This body of work has advanced in significant ways our knowledge of the relationship between gender roles and relations on one hand and population movements on the other. These studies have also contributed to a richer and more multifaceted understanding of the experiences of women migrants in China.

This chapter aims at adding to this body of work by documenting gender differentials in migration and their changes and highlighting the relationship between gender and household strategies. I argue that women are as important as men in population movements and that gender roles and relations are central for understanding household-level decision-making, strategies, and processes in relation to migration. By employing a split-household strategy, which entails gender division of labor and facilitates circular migration, peasant migrants are able to obtain the best of the origin and destination. Their double roles as

76 *Gender and household strategies*

farmers and urban workers challenge the permanent migrant paradigm that assumes that the ultimate goal of temporary, labor migration is permanent settlement at the destination (see Chapter 1).

Gender differentials in migration

Migration propensity

Migration is a selective process. Who migrates and who stays reflects differentials in the motivation to move, access to resources, knowledge, and opportunities about migration, and ability to overcome friction of distance and other obstacles. Table 5.1 summarizes the salient gender differences in migration selectivity in China. The 1990 census records 35.3 million intercounty migrants, of whom 56 percent are men and 44 percent are women. Based upon the population aged five or above in 1990, the migration rate of men and women is respectively 3.7 percent and 3.1 percent. The migrant male–female ratio – computed by dividing the male migration rate by the female migration rate – is 121, which is the sex ratio of migrants after controlling for the sex structure of the general population. Thus, for every 100 female intercounty migrants, there are 121 male intercounty migrants, indicating that men are more mobile than women. The gender imbalance is even more acute among interprovincial migrants, of whom 58.7 percent are men and 41.3 percent are women. The respective migration rates of 0.7 percent for men and 0.5 percent for women translate into a migrant male–female ratio of 142. Since interprovincial moves are generally of longer distances than intraprovincial moves, the above results support the notion that men travel longer distances than women. This notion, however, is less conclusive if one examines data on interprovincial migration, which is dealt with under "Spatial patterns."

The 2000 census shows that gender differentials have changed over time. While both male and female interprovincial migrants have increased in volume considerably, the number of female migrants has increased at a faster rate than that of male migrants. Accordingly, the distribution of interprovincial migrants by gender has become more even: 52.3 percent men and 47.7 percent women. Migration rates have become very similar between men (2.7 percent) and women (2.6 percent); and the migrant male–female ratio of 104 indicates that men and women are almost equally mobile and equally represented among interprovincial moves. Given the expectation that gender differential in mobility is smaller among short-distance moves than long-distance moves, it is conceivable that women's mobility across counties has increased to levels similar to or surpassing those of men.

Demographic characteristics

Female migrants are younger than male migrants (Table 5.1). The gender gap in age is largest among interprovincial migrants from the 2000 census. It is also

Table 5.1 Comparison between male and female migrants, 1990 and 2000 censuses

	1990 census							2000 census					
	Intercounty migrants			Interprovincial migrants							Interprovincial migrants		
	Male	Female	All	Male	Female	All					Male	Female	All
Number (million)	19.8	15.5	35.3	6.8	4.8	11.5					16.6	15.1	31.7
% of all migrants	56.0	44.0		58.7	41.3						52.3	47.7	
Rate (% of 5+ population)	3.7	3.1	3.4	0.7	0.5	1.1					2.7	2.6	2.7
Age													
Mean (years)	27.3	26.7	27.0	27.6	27.2	27.5					27.8	26.0	26.9
15–29 (%)	64.0	66.4	65.0	63.5	62.9	63.3					59.0	69.1	63.8
Marital status: married (15+) (%)	44.3	63.6	53.0	47.1	61.0	55.4					52.4	54.1	53.2
Education: senior secondary or above (6+) (%)	37.2	25.2	32.0	33.9	21.0	28.5					26.5	18.1	22.5
Origin (%)													
Streets	23.0	15.2	19.6	30.0	21.4	26.4					13.8	11.7	12.8
Towns	19.2	18.6	18.9	14.0	13.7	13.8					47.4	48.6	48.0
(Residents' committees)	–	–	–	–	–	–					7.3	7.0	7.2
(Villagers' committees)	–	–	–	–	–	–					40.1	41.5	40.8
Townships	57.8	66.2	61.5	56.1	65.0	59.8					38.8	39.8	39.2
Destination (%)													
Cities	58.4	54.3	56.6	55.2	44.7	50.8					56.9	49.5	53.4
Towns	–	–	–	–	–	–					18.7	22.2	20.4
Counties	41.6	45.8	43.4	44.8	55.3	49.2					24.4	28.3	26.2
Hukou type: agricultural (%)	45.1	58.6	50.1	51.9	59.7	55.1					77.1	80.8	78.9
Hukou location: permanent migrants (%)	53.3	55.2	54.1	44.7	49.5	46.6					13.0	13.5	13.2
Interprovincial migration % of intercounty migrants	–	–	–	34.2	30.7	32.6					–	–	40.9
Average distance (km)	–	–	–	850.5	901.3	871.5					884.5	865.4	875.4

Sources: 1990 census one percent sample; 2000 census 0.1 percent interprovincial migrant sample.

78 *Gender and household strategies*

among this group that female migrants are the youngest; they have the lowest mean age – 26 – and the highest level of concentration – 69.1 percent – in the 15–29 age range.

To illustrate the gender differentials by age, Figure 5.1 shows men's and women's age-specific inteprovincial migration rates, defined as the number of interprovincial migrants per 100 people. For both men and women, migration rates increase with age and reach their peaks at the 20–24 age group, after which they decline sharply. The decline is greater for women than for men. Given the strong tradition and social pressure for women, especially peasant women, to marry young, these results support the observation that women are most mobile before marriage (including marriage migration) and that their migration propensity declines sharply after marriage and childbirth.[39]

Several changes have occurred between the two censuses. First, as described earlier, migration rates for both men and women have increased considerably. Second, the 1990 census data show that men's migration rates are higher than women's migration rates across almost all age groups, except for ages 70 and above. By the 2000 census, however, women's migration rates have overtaken those of men at the 15–19 and 20–24 age groups. This is strong evidence that women's increase in mobility is especially concentrated in the young ages and before marriage and childbirth. Third, the 2000 census data show a dip at the 10–14 age group, which indicates that school-age children have lower mobility and suggests that having school-age children may be a deterrent to parents' migration (see also "The split-household strategy" and Chapter 7).

According to the 1990 census, the proportions married differ significantly

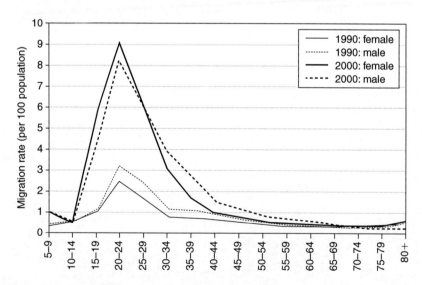

Figure 5.1 Gender differentials in age-specific interprovincial migration rate, 1990 and 2000 censuses (sources: 1990 census one percent sample; 2000 census 0.1 percent interprovincial migrant sample).

between men and women (Table 5.1). For both intercounty and interprovincial migrants, men are less likely than women to be married, reflecting in part the relative importance of "study/training" among male migrants and the large proportion of marriage migrants among female migrants (see also Figure 5.2). By the 2000 census, the gender gap in marriage rate has declined – 52.4 percent of male interprovincial migrants are married and 54.1 percent of female interprovincial migrants are married – due both to increase in marriage rate among male migrants and decrease in marriage rate among female migrants. Both are likely related to the surge in the volume of temporary migrants and the increased importance of the "industry/business" reason among these migrants (see also Table 2.2 and Figure 4.4). Specifically, many male peasant migrants are married but most of their female counterparts are single, a point I shall return to under "Maiden workers."

A large gender gap in educational attainment exists among all three groups in Table 5.1. Not only do interprovincial migrants have lower educational attainment than intercounty migrants as a whole, as reported in the 1990 census, but the former's educational attainment has declined between the two censuses, which further supports the observation about decline in migration selectivity (see also Chapter 4 and Table 4.6). The decline has occurred for both male and female interprovincial migrants, but the gender gap has narrowed. The percent point difference between male and female interprovincial migrants is 12.8 (33.9 percent minus 21 percent) according to the 1990 census and 8.4 (26.5 percent minus 18.1 percent) according to the 2000 census. In the human capital perspective, the narrowing gender gap in educational attainment has likely contributed to increased representation of women among all migrants. This supports the notion that education is an important means for women to overcome constraints on their mobility relative to men.

Spatial patterns

Male migrants are less likely than female migrants to have agricultural hukou (Table 5.1). Moreover, origin and destination data show that male migrants are less likely to come from rural origins – townships and villagers' committees – and more likely to move to cities than female migrants. These gender differences support the observation that rural–urban moves that are usually accompanied by shifts from agricultural to non-agricultural work are more highly represented by men (e.g. Wang 2000). Nonetheless, among both intercounty and interprovincial migrants in the 1990 census, women are more likely than men to be permanent migrants. This probably reflects the large proportion of women migrants moving for marriage, who are mostly rural–rural migrants and able to obtain hukou at the destination (see Tables 4.2 and 4.3).

Data from the 2000 census, again, depict some significant changes. In particular, the proportions of male and female interprovincial migrants with agricultural hukou have increased considerably, from respectively 51.9 percent to 77.1 percent and 59.7 percent to 80.8 percent. The proportions of permanent migrants, at the same time, have declined sharply, from 44.7 percent to 13 percent for men and

80 *Gender and household strategies*

from 49.5 percent to 13.5 percent for women. These results are consistent with the earlier observation that the proportion of "industry/business" migrants, who are largely of rural origin and are temporary migrants, has sharply increased between the two censuses (see Figure 4.2).

According to the 1990 census, 34.2 percent of male migrants and 30.7 percent of female migrants have undertaken interprovincial moves (Table 5.1), which supports the conventional wisdom that women move shorter distances than men. Detailed data on interprovincial migration, however, shed a somewhat different light on this notion. Specifically, according to the 1990 census, and using provincial capitals to represent origin and destination provinces, the average distance of interprovincial migration is 851 km for men and 901 km for women,[40] which suggests that female interprovincial migrants actually move longer distances, on average, than their male counterparts. A similar calculation based on the 2000 census, however, shows that the average distance of interprovincial migration is 884 km for men and 864 km for women. Changes between the two censuses indicate that among interprovincial migrants, men have increased their average migration distance, women have reduced their average migration distance, and once again support the prevailing notion that men move longer distances than women. These conflicting findings suggest that the evidence on whether men or women move longer distances is hardly conclusive.

Migration reasons

A prevailing assumption in migration research is that men move primarily for economic and work-related reasons and women move primarily for social, familial and associational reasons. In China and other countries where the status of women is persistently and significantly lower than that of men, gender discrepancy in migration motives is expected to be even more profound. Indeed, many researchers consider the work/economic versus family/social dichotomy as one of the major differentials between male and female migrants in China (e.g. Davin 1998). On the other hand, some researchers question the validity of this dichotomy and have provided evidence that a considerable proportion of women move for economic reasons and independently (Fan 1999; Wang 2000; Yang and Guo 1999). The following analysis supports the argument that economic reasons are increasingly important for women migrants (see also Chapter 8).

1990 census

Figure 5.2 compares the migration reasons of male and female intercounty migrants according to the 1990 census. Based on the economic–social dichotomy described in Figure 4.1 and excluding "study/training," a total of respectively 53.7 percent and 27.8 percent of male and female migrants move for economic reasons, and 20.1 percent and 56.6 percent move for social reasons. These statistics support the conventional wisdom about discrepancies in migration motives between men and women.

Gender and household strategies 81

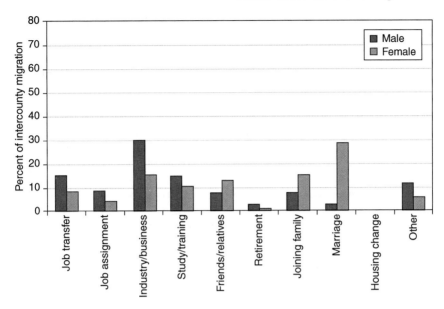

Figure 5.2 Migration reasons by gender, 1990 census (source: 1990 census one percent sample).

For male migrants, "industry/business" is the leading reason, followed by "job transfer," "study/training," and "job assignment." All the above four categories are related to one's work and to opportunities to increase one's competitiveness at work. They suggest that men's migration is especially driven by work and economic opportunities. Among female migrants, "marriage" is the leading reason, followed by "industry/business," "joining family," and "friends/relatives." The prominence of marriage as a migration reason for women suggests that they are primarily tied movers and that they are not as driven by economic opportunities as men. In Chapter 8, however, I shall examine this notion in detail and illustrate the link between marriage migration and economic motives. The relatively large proportions of female migrants moving for "industry/business" and "friends/relatives" (related to "industry/business" migrants seeking support at the destination) reasons indicate that wage work (as temporary migrants) is indeed an important motivation for female migration. On the other hand, their relatively high representation among "joining family" moves underscores the role of men as designated breadwinners. Furthermore, female migrants' low representation in "job transfer" and "job assignment" moves reflects the persistent gender gap in education and women's generally lower access to state-assigned jobs. Women's lower proportion as "study/training" migrants compared to men further points out that the former are less likely to migrate in order to pursue education.

Among interprovincial migrants based on the 1990 census, the gender gaps in the proportions of economic and social reasons are even larger (Figures 5.3 and 5.4). Economic reasons account for 64.3 percent and 27.5 percent, and

82 *Gender and household strategies*

social reasons account for 20.1 percent and 59.9 percent, of male interprovincial migrants and female interprovincial migrants respectively.

1990 and 2000 census comparisons

Between the 1990 and 2000 censuses, the relative importance of migration reasons among interprovincial migrants has changed. Among male interprovincial migrants, "industry/business" (72.7 percent) has become the overwhelming leading reason for migration, while the proportion of "job transfer" migrants (3.4 percent) has dropped sharply (Figure 5.3). This is consistent with an earlier observation (Chapter 4) that market channels for employment have rapidly gained importance while state allocation of jobs has declined in importance. The prominence of economic reasons among male interprovincial migrants has increased from 64.3 percent in the 1990 census to 78 percent in the 2000 census.

The changes among female interprovincial migrants are even more profound (Figure 5.4). The proportion of "industry/business" migrants has increased sharply from 16.6 percent in the 1990 census to 62.7 percent in the 2000 census, and it has replaced "marriage" as the most prominent reason for female interprovincial migration. The proportion of "marriage" migrants has declined from 29.1 percent in the 1990 census to 10.5 percent in the 2000 census. Economic reasons as a whole account for 64.7 percent of female interprovincial migrants, representing a substantial increase from the 27.5 percent in the 1990 census. The proportion

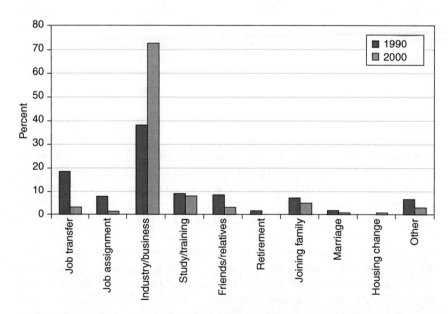

Figure 5.3 Reasons for male interprovincial migration, 1990 and 2000 censuses (sources: 1990 census one percent sample; 2000 census 0.1 percent interprovincial sample).

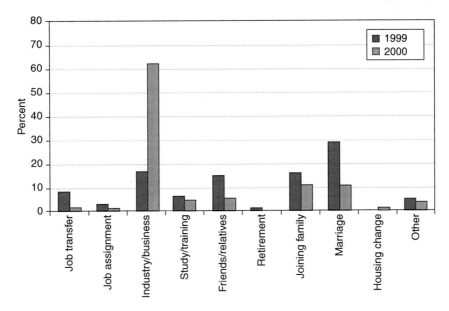

Figure 5.4 Reasons for female interprovincial migration, 1990 and 2000 censuses (sources: 1990 census one percent sample; 2000 census 0.1 percent interprovincial migrant sample).

attributable to social reasons has declined from 59.9 percent to 26.1 percent. The above is strong evidence for the rapidly increasing importance of economic motivation behind women's migration, which has likely contributed to their increased mobility between the two censuses, as observed earlier in this chapter.

The above analysis has shown that the conventional notions about gender differentials in migration are not as conclusive as they seem. In China, the gender gap in mobility has narrowed rapidly over the last two decades or so. Women are clearly and increasingly important participants in population movements. Women who undertake interprovincial migration are moving long distances and perhaps even longer distances than their male counterparts. The assumption that female migrants move primarily for social reasons rather than economic reasons is increasingly questionable (see also Chapter 8).

Gender, the rural household, and household strategies

As discussed in Chapter 1, household considerations and strategies are central to explaining migration decisions, patterns, and outcomes (Ma *et al.* 2004). In this light, differentials between male and female migration cannot be fully understood unless they are interpreted in relation to the household as a unit of analysis. The rest of this chapter is devoted to using a household perspective to explain gender differentials in migration and to identify the relationship between gender roles and relations on one hand and household strategies on the

84 *Gender and household strategies*

other. The focus here is on rural–urban labor migrants, who constitute the bulk of temporary migrants in China.

Maiden workers

Two of the most consistently reported findings on gender differentials among rural–urban labor migrants in China are, first, that men have higher migration propensities than women and, second, that female migrants are younger than and more likely to be single than male migrants. Analyses of national-level data described earlier in this chapter support these observations in general, but data on rural–urban labor migrants and temporary migrants illustrate these gender differentials even more prominently. The 1990 census documents that among temporary migrants who move from townships to cities and who select "industry/business" as their migration reason, 71.9 percent are men. The 15–24 age range accounts for 64 percent of female migrants but only 47.1 percent of male migrants. And, respectively, 58.5 percent and 48.4 percent of female and male migrants are single. According to the 2000 census, among temporary inter-provincial migrants who move from townships and village committees to cities and select "industry/business" as their migration reason, 58.7 percent are men; the 15–24 age range accounts for 53.2 percent of female migrants but only 35.1 percent of male migrants; and 51.8 percent of female migrants and 41 percent of male migrants are single.

Many other sources also report that among rural labor migrants, men are more highly represented than women and men are older and more likely to be married than women. According to a national survey conducted by the Ministry of Public Security in 1997 (*Gonganbu huzheng guanliju* 1997), which includes all temporary migrants who registered with local PSB, men account for 59.8 percent of the temporary population that are in the urban labor force. More than two-thirds (68.4 percent) of the migrants documented in the 1995 Sichuan and Anhui Household Survey are men, and more women (75.7 percent) than men (48.7 percent) are under 25 years old. Of the migrant interviewees included in the 1995 Sichuan and Anhui Interview Records, 83.8 percent are men. In that survey, the average age is 24.7 for female migrants and 31.7 for male migrants; the age group 15–24 accounts for 75.9 percent of female migrants and only 25 percent of male migrants; and 77.4 percent of female migrants and only 20.6 percent of male migrants are single. The 1998 Guangzhou survey also documents that female rural–urban migrants are younger and more likely to be single than their male counterparts.

The high representation of young, single individuals among female, rural labor migrants has given rise to the term *dagongmei* – young migrant women workers, "working women" or "working sisters" (Gaetano and Jacka 2004: 5; see Chapter 1). Based on a study of a Shenzhen factory, Ching Kwan Lee (1995; 1998) coined the term "maiden workers" to refer to young, single female migrants that are perceived to be docile workers. In Chapter 6, I shall examine the demand-side explanations – why are young, single migrant women in demand in the urban labor market? In the following, I focus on the supply side; namely, why are young, single peasant women more likely than older, married women to pursue migrant work?

As described in Chapter 1, the patriarchal sociocultural traditions that determine gender roles continue to be strong in China, especially in the countryside. These traditions explain in part why men have higher migration propensities than women. When migration of one or more family members is deemed favorable for the household's well being, men are more likely than women to be candidates because of the likelihood that the former can earn higher incomes (Knight and Song 1995) and because of the expectation that men are breadwinners and women look after the household and provide care to the young and old. This gender division of labor is further elaborated under "The split-household strategy." The phenomenon of maiden workers, at the same time, reflects the patriarchal traditions that govern the gender roles and life cycle of rural women, as illustrated in practices of education and marriage.

Education

In rural China, most young women do not pursue education beyond junior secondary and many withdraw from school after completing the primary level. The decision to discontinue at or before junior secondary may be made by themselves or by their parents, but either way it reflects the age-old view that education for the daughter is wasteful because she will eventually marry out and become a member of another household (see Chapter 1). At the same time, a large agricultural labor surplus and lack of farmland mean that many peasant women have never or seldom engaged in farming. Those in their late teens, after leaving school, may have little to do other than house chores. Having "nothing to do at home" is, indeed, a common explanation by peasant women of their pursuit of migrant work. This Sichuan woman, who left her village to work in Dongguan, Guangdong when she was 17, has two older siblings[41] and thus her labor is not needed in farming: "I am the youngest in the family and have never been needed in farming .... I was just staying at home doing nothing. So, after graduating from junior secondary I wanted to go out [to work]" (NNJYZ 1995: 247–249).

Young peasant women's pursuit of migrant work not only saves the family money by their "not eating at home" but is also a means of increasing household income and creating opportunities for others, especially male siblings. Sons are more likely than daughters to be encouraged by their parents to stay in school beyond junior secondary. In fact, it is not uncommon for young women to quit school in order to support their brothers' education. This Anhui woman left school to work before finishing junior secondary: "Three years ago, one of my older brothers was admitted to the university. This good news, however, meant additional burden on the family ... I had no options but to quit school and go to Shanghai to work" (NNJYZ 1995: 298–299).

Not only are parents and sons invested in such gender hierarchy, but daughters also contribute to continuing the traditional ideology that privileges males. This woman left school before finishing junior secondary but wants her younger brother, who was denied admission to junior secondary, to stay in school:

> The family was poor and needed money, and so I forced myself to quit school .... I planned to go back to school after I saved up some money .... Two years ago my mother became sick .... We just built a house and were in need of money. So I didn't go back to school after all. I was always a good student and have always wanted to go to school .... I am still sad that I missed the opportunity to study more .... I returned home [from migrant work] in order to persuade my brother to repeat a year [and apply to senior secondary again].
>
> NNJYZ 1995: 473–475

Even though she was unable to continue her own education, she wants to see to it that her brother furthers his education. Her story not only illustrates the traditional ideology that works against peasant women's pursuit of personal goals, but most importantly it demonstrates the gendered expectations that women are more tolerant and should sacrifice for the well being of others, especially male members, of the family. The above narratives explain why female peasant migrants are so young; they also illustrate that gender inequality in the countryside is deep-rooted and that opportunities are still very much defined by one's gender.

Marriage

The strong tradition and social pressure for peasant women to marry young persists. Marriage has two implications for female peasant migrants. First, most return to the countryside to find a marriage partner. Second, once married and especially after they have children, peasant women's propensity to pursue migrant work declines sharply (see "Demographic characteristics;" Hare 1999b; Wang 2000; Yang and Guo 1999). Both explain why young, single women are much more highly represented than older, married women among rural labor migrants.

Parents are eager for their daughters to find a husband during their "marriageable age." When peasant migrant women reach their early 20s, there is increasing pressure for them to return to get married (NNJYZ 1995: 26–28; NNJYZ 1999: 424). Although arranged marriages are officially outlawed (see also Chapter 8), it is still common for parents to pressure their daughters to get married and some parents are very involved in finding desirable marriage partners for their children (NNJYZ 1999: 214–215).

Most peasant migrant women return to the countryside for marriage (Tan 1996b). This is partly because their rural background renders them the least desirable in the urban marriage market (Chen 1999; Chiang 1999). Until recently, a child born in cities had to inherit the mother's hukou, which discouraged urban men from marrying rural women. Such marriages carried the risks of long periods of separation and of children being denied urban benefits (*China Daily* 2001). In 1998, the State Council approved guidelines permitting children to inherit either parent's hukou (Chan and Zhang 1999; Chen 1999; Davin 1998; Yu 2002), but it is unclear whether the guideline is widely adopted and whether it can offset the longstanding prejudice in the urban marriage market that disadvantages peasant women (see Chapter 3). The prevailing sentiment is that peasant women should return to the countryside to find a husband. Most are expected to return to the home village and community rather than marrying another migrant and moving to a place far from home. This 22-year-old Sichuan woman's narrative describes the above considerations:

> For sure I'll return to the countryside to find a husband. It's unrealistic to try to find a husband "outside." If you marry a migrant, you may not speak their [the husband and his family's] dialect and you will live too far from the home village. Plus, it's difficult to know what type of person they are .... As for the urban natives, only the disabled are willing to marry migrant women; very few urbanites would marry migrant women based on affection. I am going to find a husband from home .... Migrant work is not a long-term strategy. I'll return when I am 24 or 25 years old.
>
> NNJYZ 1995: 176–180

As in many other parts of the world, where married women are expected to take up care-giving roles in the home (e.g. Radcliffe 1990), marriage signals a termination of migrant work for peasant women in China. As a result, the mobility of peasant women in their late 20s and early 30s drops sharply, as illustrated earlier in this chapter. Male migrants, on the other hand, are less constrained by marriage to pursue migrant work. This gender difference is attributable to both sociocultural traditions and household strategies. As discussed in Chapter 1, the age-old inside–outside ideology is reproduced in today's countryside, where married women are tied not only to the home but also to the village and farm work (Jacka 1997). Second, as illustrated in greater detail under "The split-household strategy," temporary and circular migration is a popular strategy that entails one or more household members engaging in migrant work while others remain in the countryside. More than likely, it is the husband that migrates to work and it is the wife that stays in the village to farm and to take care of house chores and children.

The inside–outside ideology refers to not only gender roles in the household but also the perception about rural and urban lives. Married women, and even single women who are engaged, are expected to be content with village life. They are discouraged from pursuing migrant work because of the perception that the "outside" is disorderly and unpredictable whereas the village provides order and security (see also Chapter 7). This 24-year-old Anhui woman illustrates the pressure for married women to stay in the village:

> In our village, most women my age are already married. After they get married most will not pursue migrant work anymore ... there is a common

view that migrant women are loose. So, after you are engaged, your fiancé and his family will discourage you from migrant work.

NNJYZ 1995: 505–507

Her experience illustrates the traditional ideology that women are defined in relation to marriage and that marriage, even engagement, legitimizes the transfer of a woman's labor and autonomy to the husband's household (see Chapter 1). The notion that migrant women are immoral is also related to the age-old belief that women's proper place is "inside" (the home and the village).[42]

Marriage traditions in rural China – women marrying young, migrant women returning home for marriage, and staying in the village upon marriage – mean that migrant work is nothing but a short episode during peasant women's youth (Lee 1995). For the most part, marriage denotes the end of migrant and urban work for peasant women.

The split-household strategy

It is well established that the main objective of labor migration is to increase income (see also Chapters 4 and 6). As described in Chapter 4, the relative importance of economic reasons for migration, especially among temporary migrants, has increased over time. Migration for the purpose of increasing income and diversifying income sources for the household is neither new nor unique to China. How this objective shapes household strategies in migration, and how social and power relations within the household affect migration and related considerations are, however, less well understood. This is the focus of the rest of the chapter.

Migrants' intention to return

The prevailing assumption about peasant migrants' intention to stay is that they want to become permanent residents at the destination but are prevented from doing so by the hukou policy. Peasant migrants' inferior institutional status is indeed a formidable barrier to their access to urban services and the full range of urban jobs (see Chapter 6). Children of peasant migrants, for example, have limited educational opportunities in the city (see Chapter 7). The common expectation is that when offered hukou in the city, peasant migrants and their families will move there. Accordingly, a central issue in the debates about hukou reform is whether abolition of hukou will result in migrants flooding cities and runaway urban growth (Cai 2001: 335–336). Existing evidence, however, questions the notion that peasant migrants have a strong desire to settle in cities (see Chapter 1). Studies that reveal that most peasant migrants want to eventually return to their origins and do not wish to transfer their hukou to the city, even if such transfer is free, suggest that the temporary nature of the floating population cannot be fully explained by the hukou policy but is also due to household strategies aimed at increasing income and spreading risk (Cai 2000: 145; Wang 2003; Wang et al. 2002; Wen 2002; Zhu 2003; 2007).

To many migrants, the ultimate goal of migrant work is not permanent residence in the city but rather returning home to a better life later. That migrant work is not a long-term strategy is well summarized by this Sichuan man who works as a custodian in Guangdong: "Migration increases peasants' income .... But this is not a long-term strategy. I have a wife, children, and a house in the home village. I will eventually return to farm" (NNJYZ 1995: 34–36). The pursuit of (work) opportunities elsewhere with full intention to return has been commonplace in Chinese history. Focusing on late Imperial China, Skinner's work (1976) shows that it was precisely because sojourners who left to pursue their "occupational calling away from home" were expected to return that they could count on the support from the family and home community.[43]

Gender division of labor

Planning for eventual return explains the popularity of the split-household strategy – one or more household members doing migrant work while others stay at the home village to farm and to care for the young and old. This strategy allows migrants to "earn in the city and spend in the village" (Hugo 2003b) and enables them to return to improved livelihood in the countryside in the future. Gender roles, especially roles in relation to marriage, are central to the feasibility and success of this strategy.

As described earlier in this chapter, marriage restricts peasant women's mobility. Married men, on the other hand, are freer to pursue or continue migrant work, as illustrated by male migrants' older age, less concentrated age structure and greater likelihood to be married compared to female migrants (see "Gender differentials in migration"). Marriage, in fact, facilitates peasant men's migrant work. The arrival of a wife not only represents augmentation of labor resources to the husband's family, but she can also become the designated person to take care of the farmland, house chores, and children, making it possible for the husband to pursue migrant work. To this Sichuan man who works as a construction worker in Beijing, an important criterion for a wife is her ability to farm: "I had had several girl friends .... I was looking for someone who was capable and could take care of the farmland while I worked as a migrant. My wife is indeed very capable. She can take care of all the farm work by herself" (NNJYZ 1995: 164–167).

Based on the 1995 Sichuan and Anhui Interview Records, 78.1 percent of all male migrants and only 22.6 percent of all female migrants are married (Table 5.2). Split households, in which the husband does migrant work and the wife stays in the village, account for 69.1 percent of married migrant households. In only 26.2 percent of married migrant households are both spouses engaged in migrant work, and less than 5 percent are split households where the wife migrates and the husband stays in the village. Designating the spouse to take care of the farmland (59.7 percent) and children (69.4 percent) is the most

Table 5.2 Division of labor in married migrant households, 1995 Sichuan and Anhui Interview Records

	Married migrant households*
Number	
Male married migrants	125
(% of all male migrants)	78.1
Female married migrants	7
(% of all female migrants)	22.6
Types of migration arrangement (%)	
Husband is migrant; wife in village	69.1
Both husband and wife are migrants	26.2
Wife is migrant; husband in village	4.8
Person(s) taking care of farmland (%)	
Spouse	59.7
Parent(s)	22.5
Farmland subcontracted	11.6
Other	5.4
Person(s) taking care of children (%)	
Spouse in village	69.4
Parent(s) in village	13.5
Children are grown	9.0
Children are also migrants	8.1

Sources: 1995 Sichuan and Anhui Interview Records.

Note
* Households where one or both of the spouses are migrants.

popular arrangement. Only 11.6 percent of married migrant households subcontract their land to other farmers; and only 8.1 percent take their children along during migrant work. The large proportion of split households supports the observation that a division of labor along rural (wife) and urban (husband) lines is the most popular household strategy. Such division of labor is also prominent in the 1999 Gaozhou field study (see Chapter 8), accounting for about half of the married couples surveyed (Plate 5.1).

The gender division of labor that ensues is consistent with the age-old inside–outside ideology (see Chapter 1). To peasant migrant households, the "outside" refers to not only what is beyond the household but also the labor market outside the village, whereas the "inside" refers not only to the home but also to the village and agricultural work (Jacka 1997). This division of labor ties women to agricultural labor and reinforces the feminization of agriculture (e.g. Bossen 1994; Cartier 1998; Davin 1998; Jacka 1997; Zhang 1999). To women migrants, therefore, marriage is disempowering because it cuts short their wage work and disrupts their economic mobility through urban work. Men have once again become the designated breadwinner through urban wage work, on which their wives rely for improving the household's well-being.

Gender and household strategies 91

Plate 5.1 Peasant women are expected to stay to take care of the children and farmland so that their husbands can pursue off-farm migrant work (source: Author, Gaozhon, December 1999).

A much less popular split-household arrangement is one where the wife does migrant work and the husband stays in the village. Most likely, this reverse division of labor strategy is pursued because of the husband's failure in migrant work. In other words, married women migrants are but replacement migrants; men remain the preferred candidates for migrant work, as illustrated by this 32-year-old Sichuan woman's story:

> During the past several years my husband had done migrant work in mining, construction, brick factory, etc. He is unskilled and can only do manual work – work that is dirty, tiring, and dangerous .... He is impatient and has a bad temper, and he cannot tolerate the tough life of migrants .... The past several years the money he made from migrant work wasn't even enough to pay for his food, cigarettes and drinks. Even he himself admits that he is useless [*meiyong*]. For several years he didn't bring back a cent .... Fellow villagers all tease him. He feels embarrassed and doesn't want to go out anymore .... I suggested that he take care of the home so that I could try my luck outside. He said, "Don't you look down on me. Migrant work is harder than you think. Try it if you don't believe me." So, I went out and he stayed home to farm and look after the kids .... In one year I brought home 3,000 yuan. This money helped us to pay back our debt, purchase fertilizers and

pesticides, pay the children's school fees, and buy a TV set for them ... .
I still want to return to work after the Spring Festival, but my husband
doesn't want me to go ... he wants me to stay home and help him raise some
pigs ... we have been fighting about this matter.

NNJYZ 1995: 2–4

This is a vivid example of how the reverse division of labor is seen as a deviation from traditional gender role arrangements and is hotly contested. Even though this woman is happy about her economic achievements, her comment on the husband's failure in migrant work shows that she is also heavily invested in constructions of gender and the institution of marriage, even as she may simultaneously contest them and feel constrained by them. Men who stay in the village while their wives do migrant work risk being perceived as "useless" and tend to put pressure on the wives to return (see also Lou et al. 2003).

Other forms of split households

While a husband–wife split-household arrangement may be the most popular model among peasant migrant households, it is important to note that other forms of split households exist. Among these, the most common arrangements involve intergenerational collaborations. When parents are physically well and able to take care of the farmland and other house chores, their sons and daughters may be motivated to pursue migrant work in order to improve the family's economic well-being. On the other hand, a parent's sickness, especially in poor households, often compels rural men and women to look for urban work in order to pay for medical expenses.

The availability of one or more grandparents who can look after young children makes it possible for one or both parents to do migrant work. However, a common reason for peasant migrants to return is that their children are entering or have entered school (see Chapter 7). It seems that a popular view in the countryside is that grandparents are good candidates to look after pre-school-age children but older children need the parents' direct supervision. This sentiment is aptly illustrated by this 28-year-old migrant woman's comments:

When both my husband and I were outside, our child was taken care of by her grandparents. But grandparents don't have *wenhua* [culture or education]. They can only take care of the child's daily life but cannot educate her. They spoil her too much.

NNJYZ 1999: 368–369

Another 32-year-old woman expresses a similar view: "The grandma has no culture and cannot control the child". (NNJYZ 1999: 14).

In addition to intergenerational collaborations, siblings and relatives may pool their resources such that they can both take advantage of migrant work opportunities and take care of family needs in the countryside. This 41-year-old Sichuan man, who had worked for many years in Guangdong in construction and factories, decided to return when his parents became sick:

> The reason of my return is that my parents are getting old and sick and need help. I discussed with my two brothers and decided that we will take turn to come back for two years to take care of them. Since I am the oldest son, I am the first to shoulder this responsibility.
>
> NNJYZ 1999: 451

His narrative illustrates how household strategies may involve family members not living under the same roof (see Chapter 1). All the three brothers were engaged in migrant work and the decision to return was made on the basis of both economic considerations and social obligations. This is an example of collective and negotiated strategies that span two or more generations and several households. Household strategies, therefore, highlight the household as a unit of analysis but are not limited to division of labor within the physical household. Collaboration and division of labor across households are pursued when they promise to achieve mutually beneficial goals and advance the overall well-being of all households involved.

Circular migration and remittances

Migrants' maintaining strong ties with the home village and community and their continued support of the rural livelihood are essential ingredients of the split-household strategy. It is well documented that peasant migrants are highly circular (Hare 1999a; Solinger 1995). Many migrants return during the Spring Festival and during planting and harvesting seasons. Such circular migration enables them to participate in valued traditions and to provide needed labor for agriculture. For example, this 30-year-old Sichuan man is a veteran migrant and has returned to the home village frequently:

> For the last ten years, I returned home twice a year, once during harvest around August, and once during the Spring Festival .... My wife is happy that I am doing migrant work, for the income I bring home. She takes care of the farmland .... But migrant work is not a long-term strategy. After a few years I'll return home to farm or perhaps start a business.
>
> NNJYZ 1995: 6–9

Despite the impression that many rural Chinese desire to leave agriculture (Croll and Huang 1997), farming is still a major source of food and income for them, and there is evidence that most migrants value highly their access to farmland (see Chapter 1). Zhu's (2003; 2007) surveys in Fujian show that the vast majority of migrants would not give up the land even if they were given urban hukou. In addition to being a source of livelihood, farmland serves as an insurance against adversity and a security for migrants' eventual return.

The importance of agriculture to migrants and their families can be seen in how remittances are used. In Chapter 7, I shall discuss in more detail the volume and usage of remittances, but it is useful to note here that agricultural input is among the most important items for remittance spending. Almost all other major usages of remittances have to do with sustaining and improving rural living, such as house building or renovation, living expenses, education fee, repaying debts, and wedding expenses. This shows that peasant migrants are highly invested in the household and livelihood in the countryside, which also supports the argument that most migrants intend to return in the future.

Summary and conclusion

Women in China are as mobile as men. In recent decades, the magnitude of female migration has increased at a faster rate than that of male migration. Among interprovincial migrants, women move long distances, in a similar way to men. For both male and female migrants, the relative importance of economic reasons for migration has increased, considerably, over time. All of this challenges the conventional wisdom about gender differentials in migration – that men have higher migration propensities, that men's mobility is driven by economic reasons and women's movements are driven by social reasons, and that men move longer distances than women.

Despite the above, there is strong evidence that women's mobility decisions and patterns continue to be dominated by age-old patriarchal traditions. Peasant women's lack of education opportunities makes them ready candidates as labor migrants while they are young and before marriage. These maiden workers, however, are expected to return for marriage and stay in the countryside thereafter. Married women are key to the split-household strategy, which most often involves the wife staying in the village to farm and raise children and the husband pursuing migrant work. While other forms of split households exist, it is the division of labor between genders that is most popular. The popularity of this strategy highlights three important notions for conceptualizing rural–urban migration in China. First, this household strategy extends and reinforces the deep-rooted inside–outside ideology that governs gender roles in Chinese society. Second, the pursuit and persistence of the split-household strategy suggests that peasant migrants do not necessarily desire to stay in the city (see Chapters 7 and 9). Related to this are the strong ties peasant migrants maintain with the rural community and livelihood, especially via circular migration and remittances. All of this shows that peasant migrants are not merely passively responding to changing circumstances around them but are actively organizing and reorganizing resources they have access to in order to take full advantage of what both the city and the countryside can offer.

6 Migrants' experiences in the city

Introduction

The experiences of peasant migrants in the Chinese city, though varied, reflect their positions as outsiders to the urban society and labor market. This chapter focuses on the relationship between migration and labor market processes and segmentation, and peasant migrants' exclusion from urban membership and their responses to such exclusion. In the city, employment practices from recruitment and job placement to wages, benefits, and working conditions are all illustrative of the deep segmentation in the urban labor market that relegates peasant migrants to the bottom rungs. This labor market hierarchy results not only from human capital differentials but also from the institutional and social stratification that defines permanent migrants as the elite and temporary migrants as outsiders. Peasant migrants have, at the same time, developed mobility and social strategies that help them deal with the exclusion and discrimination they face in the city.

Labor market processes and segmentation

Migrants' experiences in the destination are in large part determined by their work. This section focuses on the labor market, in particular the urban labor market, and how its segmentation relates to hukou status. Emergence of an urban labor market in China has motivated migrants to enter cities; at the same time, migrants' arrival has accelerated the transformation of the urban economy and segmentation of the urban labor market (Cao 1995; Cook and Maurer-Fazio 1999; Knight and Song 1995; Liang 1999; Yang and Guo 1996). In particular, the two-track migration system, described in Chapter 4, is shaping in important ways how China's urban labor market is evolving.

Job search, the type of jobs one has, and job compensation are among the most revealing indicators of labor market processes and segmentation. However, this type of information is limited, if at all included, in census-type surveys. In what follows, I use the 1990 and 2000 censuses as well as information from the 1998 Guangzhou survey and the 1995 and 1999 Sichuan and Anhui Interview Records to elucidate the relationships between labor migration on one hand and labor market processes and practices on the other.

Job search

1998 Guangzhou survey

Table 6.1 compares the job search experiences among non-migrants, permanent migrants and temporary migrants in order to show how hukou status is related to labor market segmentation. "Income" is the leading job search criterion of all three groups but is most prominent among temporary migrants. "Nature of work unit," referring specifically to "ownership sector" (see below), is a criterion for respectively 23 percent and 18.6 percent of non-migrants and permanent migrants but only 4.7 percent of temporary migrants. These differences reinforce the notion that monetary return is an important incentive for migration but is especially important for temporary migrants, whereas non-monetary job attributes

Table 6.1 Job search, ownership sector, and job stability, 1998 Guangzhou survey

	Non-migrants	Permanent migrants	Temporary migrants
Criteria for job search (%)			
Income	40.9	45.3	62.7
Nature of work unit	18.6	23.0	4.7
Stability	24.9	13.2	18.8
Location	1.7	7.4	3.6
Benefits	4.7	5.7	3.2
Other	5.3	4.1	3.7
Information about labor market (%)			
Relatives in Guangzhou	50.8	49.0	41.2
Relatives outside Guangzhou	0.7	4.7	36.9
Advertisement	11.5	15.3	8.8
Work unit/school	22.4	11.0	0.2
Agencies in Guangzhou	4.1	6.0	4.1
Agencies outside Guangzhou	0.0	0.3	1.7
Other	9.8	13.7	6.6
How did you find this job? (%)			
Self	55.7	64.0	87.1
Recruitment	23.9	21.7	7.6
Work unit assignment	18.7	9.0	0.7
Other	1.6	5.3	4.6
Ownership sector (%)			
State-owned	43.3	57.1	16.4
Collective-owned	13.0	5.1	9.1
Foreign-invested and private-owned	22.0	20.9	55.0
Self-employed	21.7	16.9	19.4
Stability			
Number of jobs (mean)	2.2	1.9	2.5
Years at present job (mean)	9.9	3.9	2.7

Sources: 1998 Guangzhou survey; adapted from Fan (2002a) with permissions.

such as ownership sector are also valued by non-migrants and permanent migrants (see also Chapter 4). In addition, job "stability" is quite important, but "location" and "benefits" are of low importance to all three groups.

Non-migrants, permanent migrants, and temporary migrants differ in their sources of information and channels for employment (Table 6.1). Although "relatives in Guangzhou" is the leading source of labor market information for all groups, "relatives outside Guangzhou" is a prominent source of information for temporary migrants but not for permanent migrants. As discussed in Chapter 4, social networks, involving in particular fellow villagers, friends, and relatives, are the dominant source of information for temporary migrants. Significantly higher proportions of non-migrants and permanent migrants than temporary migrants use formal sources of information via "work unit/school" and "advertisement." Responses to "how did you find this job?" indicate that the majority of all three groups, especially temporary migrants, found jobs on their own, but respectively 42.6 percent and 30.7 percent of non-migrants and permanent migrants found their present jobs via "recruitment" or "work unit assignment" – channels that account for only 8.3 percent of temporary migrants. In summary, although social networks and self-initiation are important to all three groups, non-migrants and permanent migrants are more connected to institutional and organized sources and channels than are temporary migrants, who must rely on informal resources, including social networks in the origin and in the destination.

Recruitment and social networks: 1995 Sichuan and Anhui Interview Records

Migrant narratives illustrate vividly the roles of recruitment and social networks in the job search process. Contrary to the notion that rural–urban migrants are "blind drifters" (Tyson and Tyson 1996; see also Chapter 3),[44] labor migration of peasants and their incorporation into urban work do not take place haphazardly. As reflected by the marked migration streams from specific origins to specific destinations (e.g. Fan 1999) and the channeling of migrants into specific sectors and jobs (e.g. Yang and Guo 1996), the job search process of peasant migrants is far from disorderly. First, local governments, employment agencies, and employers are actively involved in recruitment activities that export workers from rural origins (Liu 1991). Recruiters do not recruit randomly, but use criteria that fulfill the demands of the migrant labor regime. This Sichuan man who works as a construction worker in Beijing explains why he was recruited:

> I was recruited by the county labor bureau to work in Beijing. Every year I work there for 11 months and return home only once. They want to recruit us because we are far from home and do not return that often.
>
> NNJYZ 1995: 164–167

98 *Migrants' experiences in the city*

Similarly, this 21-year-old Sichuan woman describes how employers coordinate with government agencies to recruit migrant workers:

> The county labor agency came to recruit workers for an electronics factory in Guangzhou .... They required junior secondary education and had specific body height requirements. They also tested my eyesight .... We paid a total of 500 yuan, including a deposit of 100 yuan to the factory, 50 yuan to the county labor agency, and transportation fee .... The county [labor agency] chartered a train car to take us to the factory.
>
> NNJYZ 1995: 174–176

In some cases, employment agencies come to migrant workers' hometowns and match them with prospective employers. This 22-year-old Anhui woman was recruited by such an agency:

> Beijing's Sanba Employment Agency came to our county to recruit workers .... The agency organized our paperwork, including temporary residence permits. We signed contracts with the agency, which specified our work hours, wage, and holidays. These rules are necessary and make us feel more secure .... I was introduced to work as a nanny for an engineer's family.
>
> NNJYZ 1995: 339–341

At both the origin and destination, therefore, government and employment agencies play an important role in recruiting the type of labor desired by employers and channel peasant migrants to specific jobs. These agencies may be associated with town or county governments in the countryside or with various levels of governments in the city.

Second, many peasant migrants do not have access to or are not aware of recruitment by local governments, agencies, and employers. Even if they do, they may not trust these sources but may prefer to rely on social networks for labor market information (Solinger 1999a: 176; Lou *et al.* 2004). For example, migrants who return home during the Spring Festival tell fellow villagers about their experiences and are then asked by prospective migrants to help them find work (see also Chapter 4 and Table 6.1). This 28-year-old Sichuan man summarizes:

> Most of the time, employment agencies cannot get you a job even after you pay the fee. Or, they'll find you low-paying jobs or jobs that are too demanding. Sometimes, the employer [via employment agencies] refuses to pay even after you've finished the work. Therefore, most people prefer to find jobs via their relatives and friends. You only need to give them a gift, or you simply owe them a favor. Relatives and friends can tell you everything about the work place, the wage, etc. Then you can decide whether you want to take the job or not.
>
> NNJYZ 1995: 132–135

Migrants' experiences in the city 99

Peasant migrants' social networks are heavily kin-, village- and gender-based. Relatives and fellow villagers inform one another how to look for jobs. For example, the Anhui woman who was first recruited by Sanba Employment Agency became a magnet and guide to other potential female migrants in the village (see earlier quote):

> In the past several years, I have brought more than 20 young women from my home village to Beijing to work as nannies .... After I returned to the village this time [during the Spring Festival], another 10 or so women asked me to take them to Beijing.
>
> NNJYZ 1995: 339–341

Peasant migrants may combine social networks with recruitment and agency-based resources. For example, fellow villagers connect migrants to employment agencies in the destination. This Anhui woman was guided by a fellow villager and succeeded in finding work through an employment agency:

> A fellow villager who worked as a nanny in Shanghai brought me there .... At first I was very scared and missed home very much .... We followed the advertisements in the bus stations and in the streets, and found a domestic work employment agency and stayed there. There are many domestic work employment agencies in Shanghai, mostly organized by street committees or resident committees. They help you connect with employers. They charge both the employer and the employee a fee .... We slept at the employment agency, either by putting a few chairs together or simply on the floor .... I was fortunate: the next day I found a job.
>
> NNJYZ 1995: 505–507

Employers may also use experienced migrants to recruit the type of labor – young, hard-working and cheap – they desire from the home village. This 22-year-old Anhui woman who works in a shoe factory in Changzhou has played such a role: "My boss asked me to introduce several fellow villagers to him. All the workers in that factory are women. Several young women from the neighboring village also want to join me" (NNJYZ 1995: 370–372). Often, migrants who have benefited from urban work are willing to be a purveyor of information for fellow villagers. Experienced migrants' sense of obligation to other villagers further reinforces the role of networks in directing migrants to specific types of work. A 46-year-old Anhui woman working as a nanny in Beijing remarks: "I have always wanted to connect fellow villagers to good jobs. When I fail I feel very bad" (NNJYZ 1995: 352–354).

Both formal recruitment and social networks foster the channeling of peasant migrants to specific destinations, sectors, and jobs and incorporate them into the migrant labor regime. Recruiters see in peasant migrants a labor force that is reputedly hard-working, tolerant, cheap, and disposable (Zhou 1998). To peasant migrants, therefore, the urban labor market consists of segments specifically

100 *Migrants' experiences in the city*

targeting their labor, rather than a range of opportunities. In addition, through social networks, new migrants replicate the work of earlier migrants, thus reinforcing the homogeneity of the migrant labor force and labor market segmentation (Fan 2001). Nevertheless, much of the existing research on migration in China has emphasized how urban labor practices and regulations limit migrants' job access, but little attention has been paid to the role of social networks in fostering labor market segmentation in the city.

Labor market segmentation: occupations and ownership sectors

Major occupation categories conventionally used in Chinese statistics include "professional and technical," "government, party and executive," "administrative and clerical," "commerce and sales," "services," "industrial production and construction," and "agriculture" (Table 6.2). The first three categories are generally considered non-menial and more prestigious. Among them, "professional and technical" is the biggest employer, whereas the workforce in "government, party and executive" and "administrative and clerical" is relatively small. Thus, in the analysis below, these three categories are collapsed into one "professional" category. The five resultant occupation categories are "professional," "commerce," "services," "industrial," and "agriculture."

Among the five categories, commerce, services, and industrial occupations have had the most drastic changes in the reform period. Commerce and services

Table 6.2 Definition of occupations

Major category	Abbreviation	Examples
Professional and technical	Professional	Scientists, engineers, physicians, finance officers, law practitioners, teachers, artists and athletes, journalists, clergy
Government, party and executive	Professional	Government officials, party officials, managers in enterprises
Administrative and clerical	Professional	Administrative workers, public security officers, postal workers
Commerce and sales	Commerce	Sales, purchasing, retail (stores or street vendors)
Services	Services	Restaurant and hotel servers, maids, hairdressers, laundry workers, custodians, gardeners, cooks, tour guides, repair workers
Industrial production and construction	Industrial	Miners, metal workers, chemical workers, garment workers, manufacturing workers, printers, machinery workers, electricians, construction workers, transportation workers
Agriculture	Agriculture	Farmers, forestry workers, fishermen

were largely restricted in the pre-reform period but have since emerged as thriving occupations. Within the industrial occupation, there has been significant growth in labor-intensive manufacturing that targets domestic consumers and the world market. Employment expansion in commerce, services, and industrial occupations has indeed accelerated the transformation of the Chinese economy. But the statuses of these occupations in China are not as established and clear-cut as in Western economies (Stinner *et al.* 1993), and there are large variations in the prestige and status of jobs within each occupation category. Nevertheless, it is generally accepted that professional and agriculture are at, respectively, the highest and lowest ends of the occupational stratum, while the statuses of commerce, services, and industrial work are between these two extremes.

Table 6.3 shows the labor force statuses and occupations of permanent and temporary migrants aged 15 or above, based on the 1990 and 2000 censuses. The majority of the migrants are working, but the proportions of temporary migrants working are much higher than those of permanent migrants. Most non-working permanent migrants are students, which reflects the prominence of the "study/training" reason among permanent migrants (see Chapter 4). Among interprovincial permanent migrants in the 2000 census, in particular, almost 40 percent are students. Among temporary migrants, however, the leading reason for not working is stay home, suggesting that these are individuals who accompany labor migrants but do not have migrant work. The stay-home proportion has, however, declined over time so that, by the 2000 census, 88.1 percent of temporary migrants are working.

Among the working migrants, there are significant contrasts in occupational attainment by hukou status. Permanent migrants are most highly represented in professional work and agriculture, and this bipolar distribution is even more distinct in the 2000 census than in the 1990 census. Peasant migrants' concentration in professional work is in part explained by their high educational attainment (see Table 4.6), while their high percentage in agriculture is related to marriage migration, which primarily involves rural–rural migrants engaged in agriculture (see Table 4.3 and Chapter 8).

Industrial work is the third-ranked occupation for permanent migrants but is a dominant leading occupation for temporary migrants. By the 2000 census, 66 percent of interprovincial temporary migrants are in industrial work, followed by commerce and services, whose proportions are respectively 12.2 percent and 8.7 percent. Temporary migrants' proportion in professional work remains small and their representation in agriculture has declined between the two censuses. These statistics reflect the important role of peasant migrants in China's pursuit of industrialization. In addition, the proportions of temporary migrants in commerce and services are significantly higher than those of permanent migrants, suggesting that the former have been a key force in the expansion of these relatively new segments of the economy. On the other hand, temporary migrants' small proportion in professional work reflects partly their low educational attainment and partly segmentation of the labor market that channels them to some occupations but excludes them from others.

Table 6.3 Occupational attainment of permanent and temporary migrants (15+), 1990 and 2000 censuses

	1990 census								2000 census		
	Intercounty migrants			Interprovincial migrants					Interprovincial migrants		
	Permanent	Temporary	Total	Permanent	Temporary	Total			Permanent	Temporary	Total
Labor force status (%)											
Working	59.9	80.3	69.2	67.6	83.3	76.0			48.4	88.1	83.0
Students	26.7	3.8	16.3	20.9	2.9	11.2			39.6	2.1	6.9
Stay home	4.4	10.5	7.2	4.2	8.8	6.6			5.3	5.5	5.5
Other non-working	9.0	5.4	7.3	7.4	5.0	6.1			6.6	4.4	4.7
Occupation (%)											
Professional*	32.4	6.2	18.5	38.1	5.2	18.7			26.5	6.9	8.4
Commerce	4.1	10.4	7.4	4.3	9.4	7.3			4.6	12.2	11.7
Services	4.5	9.5	7.1	4.4	7.5	6.2			3.4	8.7	8.3
Industrial	26.7	55.4	41.9	24.2	58.7	44.6			17.5	66.0	62.3
Agriculture	32.1	18.5	24.9	28.7	19.1	23.0			47.8	5.9	9.1

Sources: 1990 census one percent sample; 2000 census 0.1 percent interprovincial migrant sample.

Note
* See Table 6.2.

Besides occupational attainment, another important indicator of labor market segmentation is how migrants are distributed by ownership sector. This type of information is generally not included in census-type surveys. The 1998 Guangzhou survey, on the other hand, reveals large differences in ownership sectors among non-migrants, permanent migrants, and temporary migrants (Table 6.1). Both state- and collective-owned sectors are traditional socialist-type components of the state sector that have shrunk in size since the reforms. However, in large and older cities such as Guangzhou, they remain prominent. At the same time, recent reforms of SOEs and changes in the urban economy have fostered a shift of the labor force to new sectors such as the foreign-invested and private-owned sector and self-employment. These new sectors have rapidly gained prominence since the 1980s (Davis 1999). They are characterized especially by jobs in industry and services and are very important to temporary migrants, whose employment opportunities are heavily market-driven.

In the Guangzhou survey, SOEs account for 43.3 percent of non-migrants, 57.1 percent of permanent migrants but only 16.4 percent of temporary migrants. The majority of temporary migrants are in foreign-invested or private-owned enterprises. These data again support the notion that non-migrants and permanent migrants have greater access to well-established and formal labor market segments, while temporary migrants are mostly channeled to newer and less formal segments of the urban economy. In addition, roughly one in five of all non-migrants, permanent migrants, and temporary migrants is in the self-employed sector, indicating that self-employment has emerged as an important segment of Guangzhou's labor market.

Finally, the rate of job turnover is the highest among temporary migrants and lowest among non-migrants (Table 6.1). The relatively small number of jobs non-migrants have held, despite their older age (see Fan 2002a), suggests a high level of job stability among them, which is further illustrated by their long duration in the current job (averaging 9.9 years) at the time of the survey. Temporary migrants' higher job frequency and shorter duration at the present job depict a relatively high level of job turnover, which underscores not only their association with the more informal, insecure, and fluid segments of the labor market but also their highly income-driven job search approach (see Chapters 4 and 5).

Narratives from the 1995 Sichuan and Anhui Interview Records illustrate further the labor market segmentation of rural–urban migrants. Peasant migrants do not have full access to urban jobs and are overwhelmingly channeled into jobs that are dirty, dangerous, exploitative, physically demanding, and requiring long hours, that is, jobs shunned by local urbanites (Fan 2002a; Yang and Guo 1996). This Anhui man who works as a handyman in Nanjing, Jiangsu summarizes peasant migrants' jobs: "Peasant migrants are all in menial work – work that nobody else wants to do" (NNJYZ 1995: 355–357).

Migrants included in the 1995 Sichuan and Anhui Interview Records are heavily concentrated in several occupations – respectively 35.6 percent and 20.6 percent of male migrants are construction workers and manufacturing workers; and respectively 51.6 percent, 12.9 percent and 12.9 percent of female migrants are engaged in manufacturing, domestic work, and services. This type of gendered sorting is widely known among migrants. For example, this 23-year-old Sichuan woman who works in an electronics factory in Huizhou, Guangdong comments: "Most men [migrants] from our village work in construction in Jiangsu, and most women [migrants] work in Guangdong factories" (NNJYZ 1995: 26–28).

This 20-year-old Sichuan man who works as a construction worker in Shenzhen, Guangdong stresses the labor market constraints peasant migrants face:

> There isn't really a market for choosing jobs. Enterprises always prioritize recruiting employees with local hukou. We migrants don't have many options. Most of the jobs men can get via friends and relatives are in construction. It's as if there are no jobs other than construction. Last year I changed jobs twice. The only thing that hasn't changed is that wherever I went I did construction work.
>
> (NNJYZ 1995: 264–267)

His sentiment is echoed by this Sichuan woman who works in an eye-glasses factory in Shenzhen, Guangdong: "We [peasant migrants] are always the front-line production workers. Better jobs like office secretaries are always reserved for the locals" (NNJYZ 1995: 59–61).

Labor market returns

In China as in Eastern Europe, the labor market is young and characterized by an uneasy blend of socialist practices and new market mechanisms (e.g. Domanski 1990; Kornai 1997: 200–202; Stark 1986; Szelenyi and Kostello 1996). In pre-reform China, a labor market was virtually nonexistent, job mobility was low, underemployment and hidden unemployment were rampant, and wages were kept low and did not reflect performance. Urbanites received from the state, primarily through the work unit, plentiful benefits, including housing, education, and health care subsidies, which substantially compensated for the prevailing low wages. Since the 1980s, performance-based hiring and compensation has increasingly characterized the emerging urban labor market, and wage has become an increasingly useful indicator of labor market returns. Labor market segmentation, discussed earlier, that relegates rural–urban migrants to low-paying jobs and to non-state sectors is directly translated into low wages. According to the 1998 Guangzhou survey, permanent migrants have the highest mean monthly income, followed by non-migrants, and temporary migrants have the lowest mean monthly income. While these differentials are partly functions of the educational attainment and occupations of the three groups, statistical analysis shows that after controlling for education and occupation, hukou status has significant, independent effects on income returns (Fan 2001).

The extent and types of employer-provided benefits vary considerably among the three groups in the 1998 Guangzhou survey. The majority of non-migrants and permanent migrants but only very small proportions of temporary migrants have access to health insurance and retirement benefits through their employers. The discrepancies are due partly to non-migrants and permanent migrants' higher representations in the state sector, where work units continue to provide generous benefits and subsidies to employees. Temporary migrants, on the other hand, concentrate in jobs that do not provide benefits. Even when peasant migrants work in SOEs, they are usually hired as contract workers without access to benefits (see also "Exploitative working conditions").

Discrimination and exploitation

The migrant labor regime is one that is grounded on maximization of productivity by extracting as much labor as possible from migrant workers (see Chapter 1). This demand-side explanation is exhibited through labor market practices that are discriminatory and exploitative. At the same time, peasant migrants' main objective of monetary gain and their plan to eventually return to the countryside mean that job tenure, loyalty, and career development are of little importance (see Chapters 5 and 7).

Discriminatory hiring

Age and physical attributes are among the most important criteria, often stated explicitly in advertisements, used by urban employers to recruit workers (Tan 1996b; Plate 6.1). Maiden workers (see Chapter 5) are attractive to urban employers because youth correlates with physical ability, short tenure, low seniority, and low wages and a status of single fits the demand for long hours and minimum disruption. Employers are openly ageist and discriminate against married women. For example, a Shenzhen electronics factory that Tam (2000) studied refuses to renew contracts with workers older than 20 because the management believes that by that age the worker's eyesight has already deteriorated. Most of the manufacturing jobs available to female migrants in the 1995 and 1999 Sichuan and Anhui Interview Records involve labor-intensive work in factories, such as those in electronics and garments. Women in their late 20s and early 30s are already disadvantaged and have greater difficulty finding factory work than younger migrants. For example, a 29-year-old Sichuan woman was not certain if she could find work: "I wasn't sure if factories would hire me since I was older (than other migrant women). But I looked young and most people thought that I was only 25 or 26 years old" (NNJYZ 1995: 65–68).

Service work in hotels or restaurants likewise strongly prefers young, single women for their youthful and attractive appearance. To urban employers, single women who have no family responsibilities are more desirable than married women (Lee 1995). An informant in Guangdong summarizes this preference: "It's difficult for married women to find work. Employers prefer young, single workers. When women have a family they need days off and employers

Plate 6.1 Job advertisements in cities usually specify requirements of hukou, age, sex, and level of education (source: Author, employment agency in Beijing, July 2001).

don't like that" (author's interview, Gaozhou, Guangdong, December 1999). Older migrant women are often relegated to jobs that pay less than factory work. These include domestic work and agricultural work in urban outskirts. This 41-year-old woman who works as a vegetable farm worker, for example, is deemed too old for factory work:

> Vegetable farming is hard and poorly paid work. Most people prefer not to do such work. But I have no choice. I am too old for factory work. Factories won't hire me. The minute they see my identity card [and know how old I am], they turn me away.
>
> NNJYZ 1995: 124–126

Related to age, physical attributes such as stature are also important criteria. For example, this 32-year-old Sichuan woman describes her experience:

> I was worried that I would not be able to find a job because of my age .... But everybody said that I looked young and encouraged me to try .... I was hired [by a machinery parts factory] only on probation because the production manager was concerned that I was small and thought that I could not manage the hard work.
>
> NNJYZ 1995: 2–4

It is well known among peasant migrants that the types of work they can get are most likely physically demanding. Age and physical ability are, therefore, among the reasons for some migrants to return home. This man in his late 40s and worked as a construction worker in Guangdong comments: "I worked for three years and decided to return. I was over 40 years old and could not take on tough physical jobs any more." (NNJYZ 1999: 349).

Gender roles and stereotypes are prominently exhibited in migrant worker hiring practices. Like young, female workers in export-processing zones in other parts of the world, maiden workers in China constitute an attractive magnet to global capitalist investors (e.g. Chant and Radcliffe 1992; Cheng and Hsiung 1992). These young, single migrant women are perceived and expected to be docile, detailed, able to handle delicate work, easier to control, and more complacent than male workers and older women (e.g. Lee 1995; Tam 2000). These stereotypes are well known among migrants as well, as illustrated by this male migrant who worked in Shanghai: "It is easier for women [migrants] to find jobs. Factories like to hire them because they do not expect a high pay. Male migrants have too many requests and expect to be paid more" (NNJYZ 1999: 58–60).

Exploitative working conditions

In the migrant labor regime, maximization of productivity is also realized by suppressing cost and benefits, long hours of work, minimal disruption to production, and disciplinary regulations. Low cost is in part achieved by the widespread exploitation of peasants' labor, which is legitimized and made possible by their inferior hukou positions (Knight *et al.* 1999; Solinger 1999b). The story of this Sichuanese woman, who works in a toy factory in Dongguan, Guangdong, is typical among migrants in the Sichuan and Anhui Interview Records:

> I work eight hours every day, including weekends, and get one yuan extra for four hours of overtime. We are not allowed sick leave. Even when I am sick, I must still go to work and would not let my boss know because I am afraid to. We eat and sleep at the factory – 12 to a room ... it's very crowded ... . We are paid by the piece; I don't know how they calculate my wage – the factory doesn't tell us how our wages are calculated ... . I make about 200 yuan a month, and I pay 60 yuan for my meals in the factory.
>
> NNJYZ 1995: 85–86

Peasant migrants' wages are low by urban standards, in part due to the large supply of rural labor and in part because of labor market segmentation that crowds migrants into low-paying jobs. This 28-year-old man from Sichuan who works as a janitor in Fuzhou, Fujian explains the relationship between migrants' low wages and the large supply of rural labor:

> There are too many Sichuan migrants. Employers are constantly and eagerly dampening our wages because of that. They are not afraid of not finding migrants to fill the job. If you don't take the job, there is always someone else who is willing to do it. Some fellow villagers get into fights competing for jobs.
>
> NNJYZ 1995: 132–135

Migrants share the sentiment that when faced with exploitation there is little they can do. This 20-year-old seamstress who works in a private-owned apparel factory in Jiangsu provides a succinct analysis:

> I work 17 hours a day. I start working early in the morning and go to bed in the evening only when I am extremely tired . . . . Every job is the same. The employers make big money, and what we make is not even worth their change . . . . These bosses always want to find ways to squeeze our pay. Sometimes they say that the sales are not good and they'll lower our wages. Nobody in the government, either at home or in the city, speaks for us. If governments at both ends provide specific agencies to offer us support, then we can have a place to turn to when cheated by employers . . . . We know we are exploited, but we have no choice but to work as hard as possible in order to make more money.
>
> NNJYZ 1995: 470–471

Another seamstress working in Jiangsu gives a similar account:

> The employer initially offered us a rate of 0.6 yuan per jacket, but reduced it to 0.5 yuan when he found out that we could work fast. In order to make as much money as possible, we work approximately 15 hours a day. Three of us can make 50 yuan a day. But bosses are all very mean. For each jacket their net profit is four to five yuan . . . . The local government rarely speaks for us migrant workers.
>
> NNJYZ 1995: 473–475

Unlike "regular employees" (*zhengshi zhigong*), who are permanent employees and who enjoy benefits such as health insurance and housing subsidies, peasant migrants are usually considered temporary workers (*linshigong*) or contract workers (*hetonggong*), even though they work full time and long hours. Such categorization is largely based on hukou status and no doubt helps employers reduce cost. Even in SOEs, most peasant migrants are hired as contract workers and are denied benefits that permanent employees are entitled to (Maurer-Fazio 1995; Solinger 1999b). A man working in a state-owned construction enterprise in Hefei, Anhui comments:

> Our work is dangerous and difficult. I work at least 12 hours a day and make only 11 yuan. There is practically no compensation for work-related

injuries. Our contract states that compensation is provided only for work-related deaths and lifetime disabilities. Even in those cases, the compensation is a one-time benefit and is low – maybe 10,000 to 20,000 yuan. But you are completely on our own if *only* [author's emphasis] your arms or eyes are injured ... so, we are very careful not to get injured.

NNJYZ 1995: 411–413

Not only do employers require peasant migrants to work long hours, they also go to great lengths to prevent disruption of work. One popular strategy is to deduct or withhold workers' wages and bonuses. A Sichuan woman working in Shenzhen describes this practice:

> We get a one-and-a-half month's wages as bonus if we don't take off at all during the year. If we return home for the Spring Festival then we'll lose the bulk of that bonus.
>
> NNJYZ 1995: 59–61

Denial of sick leave and paying by the piece also discourage workers from taking leaves. An Anhui woman who works in a shoe factory in Luoyang, Jiangsu describes her experience:

> We are paid by the piece; so the more we work the more we will earn .... We work 12–13 hours a day, and make about 200 yuan a month .... Two years ago, one of my fingers was crushed by the machine. The factory paid the medical expenses but also stopped paying me wages. In order to make more money, I rested for only one week before going to work again.
>
> NNJYZ 1995: 317–319

Driven by the desire to increase income and constrained by the low wage rate, peasant migrants tend to work as many hours as possible. A 23-year-old Anhui woman working in a food-processing factory in Tianjin describes this sentiment:

> The factory has three eight-hour shifts. Our wages are determined by the number of pieces we complete .... Everybody wants to work more, even a few minutes more, in order to make more money.
>
> NNJYZ 1995: 294–296

Another means to keep workers working is to provide living spaces, in addition to production spaces, a strategy favored by many foreign and domestic investors. They require migrant workers to eat and sleep in the factory dorms or in employers' homes, and often deduct room and board from the wages. This requirement makes it easy for management to impose disciplinary regulations on the workers, facilitates employees' working long hours, and isolates migrants from the world outside the work place.

Peasant migrants as outsiders

Peasant migrant workers are brought in to urban areas because of their cheap labor. In the city, they are institutionally and socially categorized as outsiders and are both physically and socially segregated from urbanites.

Outside labor

As temporary migrants, peasant workers in the city are seen and treated as outsiders. Numerous rules and regulations constantly remind them that their home is in the countryside and that they do not belong to and are not members of the city. Peasant migrants must apply for and carry all kinds of permits, which are both institutional and symbolic markers of barriers against their membership in urban society. A worker in a plastics factory in Shenzhen describes:

> Before I started working in the factory, I had to obtain an identity card, a health certificate, and a "single" [unmarried] certificate. At the factory, I obtained a temporary residence permit, a factory permit, and another health certificate. The factory deducted the permit fees from my wage .... If you are caught in the streets without a temporary residence permit, the PSB can fine you 300 to 500 yuan.
>
> NNJYZ 1995: 94–96

Even though migrants may choose not to apply for the necessary permits, they do so at the risk of being fined and arrested.

Institutional inferiority of migrant workers is also translated into and reinforced by physical segregation. Migrant workers who live in factory dorms or employers' homes are in essence entrapped within an environment designed solely to exploit their labor. A 20-year-old Anhui woman working in a garment factory in Changzhou, Jiangsu describes her working condition:

> We live and eat in the employer's home, and we work 12 hours a day [on two shifts] .... We have less than ten minutes for meals ... we must take turn to eat because the machine doesn't stop .... After work we mostly just sleep, eat, and do laundry.
>
> NNJYZ 1995: 470–471

Like her, many migrants indicate that they stay in employer-provided living spaces, typically six to 12 people to a room, and that after work they are too tired to go out. For example, this Anhui woman who works in a shoe factory in Changzhou describes:

> We work 12 hours a day, and the work place is very dirty, with soot and cotton fragments flying everywhere .... We all live in the employer's

home .... Because of the long working hours, after work we just want to go to bed and do nothing else.

NNJYZ 1995: 370–372

The blurred boundary between peasant migrants' work and living spaces stands in stark contrast to the clear demarcation between their work place and the world outside. Most factories have gates, which in addition to providing security also symbolize the impermeability between migrant workers and urban society. A migrant woman in Dongguan states:

I seldom leave the dorm; the factory has a security guard; everyone of us has a factory permit – without the permit you cannot get in .... After work I am usually too tired and go to bed early.

NNJYZ 1995: 85–86

Social exclusion

The notion of outsiders is also rooted in the cultural practices of membership in Chinese society (Fan 2002a). The concepts and vocabularies of migration connote one's membership with respect to the host community rather than the act of migration. For example, the literal translations of "migrant" – *qianyi zhe* or *qianyi di ren* – are rarely used. In their places are terms emphasizing migrants' lack of a permanent home, for example, the floating population (*liudong renkou*), and their lack of membership in the host community, for example, "people from outside" (*waidiren, wailairen,* or *wailai renkou*) (Duan 1998; see also Chapter 2). Similarly, the terms *dagong, dagongmei, dagongzai* and *nongmingong* (peasant workers) make reference immediately to peasant migrants' rural origin and their reason to be in the city – to provide labor (see also Chapters 1, 7 and 9). Their meaning to the Chinese city is mainly tied to their membership to a hardworking, tolerant, cheap, and disposable labor force (Knight *et al.* 1999; Zhou 1998). Urbanites, on the other hand, are the "insiders" or the natives – *bendiren* – that belong to the city. Differences in dialects, cuisine, and regional cultures further segregate peasant migrants from urbanites (Duan 1998).

Existing evidence so far shows that the prospect of peasant migrants' assimilation in cities is poor (Chan 1996; Fan 2002a; Solinger 1995). Comment after comment from peasant migrants illustrates that they are socially excluded and that the discrimination they face in the city is direct and blatant: "When we *waidiren* walk or bike we give way to the locals. They'll give you a hard time if you're not careful". (NNJYZ 1995: 267–270). "We don't interact with the locals in Guangzhou ... we don't understand their dialect. They look down upon us". (NNJYZ 1995: 234–235).

This 30-year-old woman who works in Shanghai describes the daily exclusion she faces:

> The most distinct impression I had during my migrant work was the discrimination by local Shanghainese toward migrants … . One time, when I was about seven or eight months pregnant, I was riding the bus with my four-year-old son. The passengers knew we were migrants [since urban people are not allowed to have more than one child]. No one gave their seats to us. At the next stop, a Shanghainese woman with an older child got on to the bus. The ticket seller immediately urged other passengers to give up their seats for her. I was outraged. We got off the bus and never took the bus again.
>
> NNJYZ 1999: 13–14

Peasant migrants' experiences of exclusion draw not only from interactions with urban residents but also from interactions with the urban government, as described by this migrant:

> When we have problems we would never go to the local government or PSB to get help. Once they notice your accent they will ignore you and protect the interests of the locals. Regardless of whether you are right or wrong, you are doomed.
>
> NNJYZ 1995: 132–135

Impacts on urban areas

Peasant migrants' impacts on urban areas are hotly debated (Jiao 2002). Migrant labor is seen as important for stimulating urban economy and facilitating the expansion of industries and services (Cao 1995; Zhong and Gu 2000). By engaging in low-paying, manual, and services types of jobs, peasant migrants are enabling urbanites to specialize in more prestigious jobs. Some urban jobs, such as domestic work, are almost exclusively taken up by migrants (Huang 1999). Migrants in cities also generate consumption, which in turn provides employment for others. Based on a survey in the mid-1990s, Zhong and Gu (2000) report that migrants' consumption accounts for more than half of the total retail consumption in Wuhan. Peasant migrants are, no doubt, a major force in shaping the urban economy.

By augmenting labor in urban areas, peasant migrants are indirectly suppressing wage increase in cities (Qiu *et al.* 2004). Based on estimates of cost ratio between local labor and migrant labor – 5 to 1 in Shanghai and 1.8 to 1 in Nanjing – Cai (2002: 218) argues that peasant migrants' productivity is higher than that of local urban labor. Moreover, the large agricultural labor surplus supports a continued supply of new, young, and cheap migrants for cities (Yang and Ding 2005). Peasant migrants are, therefore, a source of "perpetually young" labor for urban development and are especially relevant for cities experiencing or anticipated to experience population aging (Wang *et al.* 2002). Recent labor shortage in the Pearl River Delta and other areas specialized in labor-intensive manufacturing, however, suggests that peasant migrants are increasingly selective in urban work (Jian and Zhang 2005; see also Chapter 9).

Despite peasant migrants' contributions to the urban economy, public and official evaluations of them are mixed. Peasant migrants are criticized for overloading urban infrastructure such as transportation and housing, creating chaos in urban management, violating the birth-control policy, engaging in criminal activities, and spreading sexually transmitted diseases (Cao 1995; Messner *et al.* 2007; Solinger 1999a; Yang 2006; Yang *et al.* 2007; Zhong and Gu 2000). Migrants are also blamed for exacerbating urban unemployment, especially given the increase of laid-off (*xiagang*) urban employees from SOEs (Jiao 2002; Yang and Ding 2005). This criticism prompted many cities to tighten migration control in the mid-1990s (Cai 2002: 236), although scholars have shown that migrant labor and urban local labor are complementary rather than competitive (Wang *et al.* 2002; Zhong 2000: 208). Specifically, workers laid off from SOEs are more experienced and skilled than the average peasant migrant. A report by Cai and Wang, cited in Jiao (2002), concludes that the replacement ratio between the two types of labor is only 0.1. Some researchers, however, warn that the competition between peasant migrants and laid-off urban workers in cities may have increased (Cai 2002: 218). In general, a consensus among researchers is that the positive impacts of rural–urban migrants on urban areas outweigh their negative impacts (Jiao 2002).

Migrants' responses

The outsider experiences of peasant migrants have, inevitably, affected their perspectives toward migrant work, their intention to stay and their strategies.

Planning to return

Viewing the city as primarily a place of employment and for monetary gain, peasant migrants are unlikely to have long-term plans to stay in any one destination (Yang 2000a), as illustrated by this construction worker in Shenzhen: "We [migrant workers] are here to make money, and we will try to be as tolerant as possible. We are tough and will go anywhere and will adjust under almost any situation". (NNJYZ 1995: 225–227).

The lack of job security and prospect for career development discourages loyalty and contributes to high job turnover (see Table 6.1), as described by another construction worker in Shenzhen: "Bosses can fire you or fine you as they please. I will immediately switch to a new job if it offers higher wage". (NNJYZ 1995: 264–267).

Even though some peasant migrants have worked in the city for an extended period of time, their lack of urban hukou means that they have few options other than eventually returning to the countryside. A man from Sichuan working in a repair shop in Guangzhou comments:

Will I be staying in Guangdong for good? Absolutely not. We don't have [local] hukou there, which makes life very difficult. My children will be going to school soon. Without a hukou in Guangzhou we cannot afford sending them to school there.

NNJYZ 1995: 30–34

Discrimination and social exclusion have, in addition, discouraged peasant migrants from staying. This 31-year-old woman and her husband decided to return to their home village after doing migrant work in Shanghai and Wuhan: "The most important reason for us to return is that we felt we were looked down upon by urban people. My husband could not tolerate such disrespect". (NNJYZ 1999: 192–193). The above sentiment is consistent with observations made in Chapters 5 and 7. Specifically, rather than anticipating staying put in the city, peasant migrants are motivated to pursue a split-household strategy, engage in circular migration, and plan to eventually return to the countryside.

Migrant communities

Given the labor market segmentation and social exclusion that peasant migrants in cities experience, it is not unreasonable to conclude that they lead a passive and vulnerable existence with little hope. As discussed in Chapters 5 and 7, however, many peasant migrants are actively engaged in building a promising future in the countryside. In addition, there is evidence that some migrants in urban areas are carving out social and economic spaces to advance their well-being and to even enable a long-term presence there.

The networks that facilitate migration and job search remain strong in the city. For example, the pioneer migrant from Anhui who brought more than 20 women to Beijing has maintained close contact with most of her fellow villagers (see quotes under "Recruitment and social networks"):

Every Sunday I go out with 15 or 16 *laoxiang* [or *tongxiang*],[45] all of whom have worked as nannies in Beijing. We go to the park, shopping or the movies. It's a lot of fun .... Presently, some of my *laoxiang* continue to work as nannies, some are washing dishes in restaurants, and some have become food vendors. Three women married *getihu* [private entrepreneurs] or farmers in Beijing's outskirts. One woman died from a gas leakage while working in a restaurant. Her family received only 3,000 yuan in compensation. I have lost contact with three other women.

NNJYZ 1995: 339–341

Among some women migrants, a sense of sisterhood and mutual responsibility has developed, further reinforcing gendered networks. Such networks can also develop into something resembling a collective front for negotiating options in the labor market. Rather than functioning as an individual job seeker, the migrant may become part of a group that negotiates through the complex urban labor market as a single unit. After working in Shanghai as a nanny for four years, this Anhui woman contemplates changing jobs:

> For four years, I had gone to Shanghai every year to work as a nanny and I had been with seven different families. Last year, after the Spring Festival, three other women from my home village and I decided to try out factory work. So we traveled to Nanjing together, and then took the train to Changzhou and entered a private enterprise. After nine days, we all felt that factory work was too exhausting, and so the four of us returned to Shanghai and continued working as nannies.
> NNJYZ 1995: 511–513

Furthermore, social networks bind migrant women together so that many are part of a native place-based community that serves social functions not provided by urban natives. This Anhui woman who works in a food-processing factory in Tianjin describes:

> Workers in that factory came from eight different provinces, including many *tongxiang* from Anhui. We take care of one another and do things together – like taking pictures and going to karaoke bars. Every week all the *tongxiang* have a party, and sometimes we invite friends from other provinces as well. In the four years I have been at the factory, no one has ever harassed me. Even though I could have made more money working in another factory, I do not want to give up my job. It's fine that I make less money than I could have. Fellow workers in that factory are mostly in their late teens and early twenties and are friendly and full of life. As a result, factory life is a bit like campus life. We seldom interact with the Tianjin natives.
> NNJYZ 1995: 294–296

There is no question that networks – especially those involving *tongxiang* and fellow villagers – are crucial for peasant migrants to overcome the dreaded migration journey, to find work, and to enjoy social and community life. In the above examples, gender influences the formation and composition of such networks, with the ironic effect of both homogenizing migrants' experiences and occupations and providing a basis for companionship and support.

Native-place based networks can, likewise, be bases for thriving migrant communities that make it possible for some peasant migrants to stay in the city permanently. The most notable example of these successful communities is Zhejiang Village in Beijing – a sizable settlement for the residential, industrial, and business activities of migrants primarily from Wenzhou, Zhejiang. Relying on social and kinship networks (across the country) and shrewd business sense, these migrants have managed to take roots in the city despite repeated efforts by the city government to bulldoze their compounds (Ma and Xiang 1998; Xiang 2005). However, the experience of Zhejiang Village is not frequently repeated in other cities. Most peasant migrants are unable to establish economic and settlement niches that enable their permanent stay in the city.

Summary and conclusion

This chapter has focused on the labor market and social positions of peasant migrants in cities and the processes that led to these positions. The urban labor market in China is deeply segmented between temporary migrants on one hand and permanent migrants and urban natives on the other. This segmentation reflects not only human-capital differentials but most importantly the roles of recruitment and migrant networks in channeling peasants to jobs that are physically demanding, low-paying, and offering no benefits. Marked by discriminatory employment practices and exploitative working conditions, the migrant labor regime is legitimized and made possible by the developmentalist state's embracing of productivity goals and its use of the hukou system to crowd peasant migrants into peripheral segments. The state not only disengages itself from peasants' livelihood in the countryside (see Chapter 1) but also ignores migrant labor interests in the city. Although existing theories offer useful guides for interpreting the existence of labor market segments, the roles of the state and its institutional tools are the key to understanding labor market processes and segmentation in China cities.

To the Chinese city, the value of peasant migrants is primarily in their cheap, tolerant and disposable labor. They are socially excluded, discriminated against, and at the bottom of the social and economic hierarchy. Despite the important role peasant migrants play in fostering urban and industrial development, they are persistently seen as outside labor and are relegated to inferior, quasi-illegal, statuses in the city. Thus, by shaping migration and labor market processes, the state has also engendered a socioeconomic order in the city clearly stratified between urban and rural origins. Channeled to the most informal, insecure and fluid labor market segments and facing a dismal future in the urban economy and society, peasant migrants see themselves and operate as outsiders as well, employ strategies driven by short-term income gains, and rely on origin-based rather than urban resources. All of this sheds light on peasant migrants' intention to stay or return and their impacts on rural areas.

7 Impacts of migration on rural areas

Introduction

The impacts of labor out-migration on rural areas in China are multifaceted but are generally not well understood, partly because most studies consider migration as an outcome of, rather than an explanation for, change. Specifically, researchers have been more interested in explaining the processes and patterns of migration than documenting its impacts and consequences (Zhou 2002). And, among studies that do address the impacts of migration, most focus on the destination rather than the origin (see Chapter 6).

Nevertheless, migration has important implications for the countryside. More than 60 percent of China's population are rural. Rural–urban inequality is large and has been increasing over the past two decades or so (Sicular *et al.* 2007).[46] Whether migration reduces or exacerbates rural–urban inequality is, therefore, increasingly a burning question. What is clear is that rural–urban migration is, and will continue to be, sizable (Cai 2001; see Chapter 9). As discussed in Chapter 5, peasant migrants are highly circular and they maintain strong ties with their places and communities of origin. A significant proportion of migrants have, indeed, returned. Whether peasant migrants will stay in the city or return to the countryside is central to the debate on hukou reforms and migration policy (see Chapter 3).

This chapter focuses on the impacts of migration on the countryside. It begins by discussing whether labor out-migration has positive or negative impacts on the economic development of rural areas. I review major findings on how migration affects rural income, productivity, agricultural production and labor, and poverty and inequality. The second section focuses on return migration. Lastly, the chapter highlights the ways in which peasant migrants have fostered social change in their household and home community.

Economic impacts

Migrant income and remittances

Most researchers consider labor out-migration as a positive factor in raising income, diversifying income sources, increasing productivity, and alleviating

118 *Impacts of migration on rural areas*

poverty in rural China (Cai 2001: 329; CASS 2003: 54; Davin 1999; Fan 1997; Hare 1999a). There are three common methods to evaluate and report income from migration. First, rural migrants' total income is counted as part of total rural income. Based on a survey in 1994, Cai (2001: 344) reports that peasant migrants' income accounts for 18 percent of total rural income.[47] Annual surveys by NBS have consistently identified migrant income as the leading reason for the increase in rural income (CASS 2003; 2004).[48] They show that migrant income accounted for respectively 9.25 percent and 13.2 percent of total rural income in 1999 and 2003 (CASS 2004: 126).[49] Using the 2000 census and county-level socioeconomic data, Ma *et al.* (2004) find that a 1 percent increase in migration rate results in a 4.6 percent increase in rural income.

Second, income from peasant migrants can be expressed as a proportion of migrant households' total income. Unlike the first method, this approach focuses only on migrant households and excludes other households. Based on the 1995 Sichuan and Anhui Household Survey, Du and Bai (1997: 130) find that migrant income accounts for, respectively, 42.3 percent and 38.6 percent of the total income of migrant households.

Third, remittance is migrant income less expense and refers to the actual amount the household receives from migrants. The 1995 Sichuan and Anhui Household Survey reports that remittances account for respectively 19.9 percent and 23.3 percent of the total income of migrant households (Du and Bai 1997: 131). Average remittance per migrant household is 1,874 yuan for Sichuan and 2,090 yuan for Anhui. The 1999 Sichuan and Anhui Household Survey reports that migrants' average remittance is 2,853 yuan. Moreover, the survey shows that average remittance by interprovincial migrants is higher than that by intraprovincial migrants by 363 yuan, suggesting that higher income is an incentive for rural migrants to opt for most distant destinations.[50] A survey in Jinan in 1995 reports that 82 percent of migrants send home remittances, averaging 1,776 yuan per year (Cai 2000: 198). Another survey conducted in 1995 by the Chinese Academy of Social Sciences (CASS) documents that remittances represent 25.1 percent of migrant households' total income (Li 1999).[51] Lian (2002) observes that remittances to the central and western region provinces increased from the mid-1990s and have become a significant source of revenue to these areas.[52] Estimates of remittance cited above and in other studies are quite consistent and indicate that between the mid- and late 1990s an average peasant migrant worker sends home about 2,000–3,000 yuan annually and that remittances account for 20 percent to 30 percent of total household income. Using an estimate of 50 million migrant workers, Goodkind and West (2002) estimate that the annual total amount of remittances is 100–150 billion yuan. Based on 80 million migrants and average remittance of 2,000 yuan, Cai (2000: 198) arrives at total remittance of 160 billion yuan in 1995, an amount equivalent to almost four times the government's subsidies for agriculture that year.

Surveys have identified, consistently, the major items for which remittances are used. The 1995 Sichuan and Anhui Interview Records documents that remittances are used, in descending order, for house building or renovation,

Impacts of migration on rural areas 119

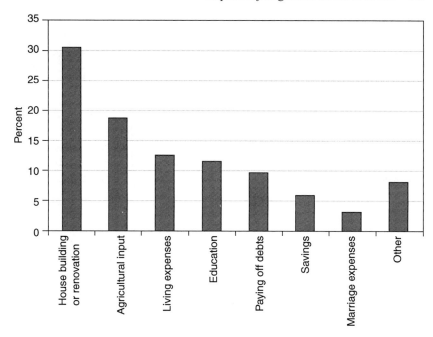

Figure 7.1 Uses of remittances, 1995 Sichuan and Anhui Interview Records.

agricultural input, living expenses, education, paying off debts, savings, and marriage expenses (Fan 2004b; Figure 7.1). These findings are similar to those from the 1999 Sichuan and Anhui Interview Records (NNJYZ 1999; Wang and Fan 2006). Murphy's (2002: 91) survey in Wanzai county in Jiangxi, conducted in 1996–1997, shows that house-building is the highest priority in remittance spending, followed by educating and raising children, daily expenses, agricultural input, paying off debts, savings, and marriage expenses. Only 2 percent of the 190 responses select "investment in business" as a spending category. These studies suggest that remittances are primarily used to fund household "projects" (such as house-building), maintain regular household activities (such as living expenses and agricultural input), support household members (such as education), and lift the household out of financial difficulties (paying off debts), rather than engage in new activities. Decisions on how to use the remittances reflect directly the relationship between peasant migrants and their places of origin, a theme addressed in Chapters 5 and 6.

Migrant work is widely perceived as the best, perhaps only, means for peasant households to improve their economic well-being. As discussed in Chapter 1, the post-Mao state has largely disengaged itself from the livelihood of the peasants, who must rely on themselves to make ends meet. Families that are in poverty and those in debt are especially compelled to pursue migrant work in order to survive, as commented by this 22-year-old woman: "There is no way out other than migrant work" (NNJYZ 1995: 517–518).

For many rural Chinese, their dream is to build a new house or to renovate their old house. Agricultural input is another important expense item. This Sichuan woman's commentary illustrates vividly the consistent finding that house-building, living expenses, and agricultural input account for the bulk of migrants' remittances:

> Our [peasants'] main goal is to build a house. You see, our family is still living in an old house, but other families [in this village] have already built new houses. My children are growing fast. We must get the house built. There are many other expenses: my youngest boy is still in school, and we need to buy fertilizer, seeds, and herbicide. We are poor ... farming is no way to make money .... I have no alternatives but to go out.
>
> NNJYZ 1995: 124–126

The impacts of migrant remittances are tangible and visible through, for example, new or renovated houses built by migrants and the clothes they wear, as commented by this 44-year-old villager: "Migrant households have higher incomes. They all have built new houses, eat better foods and wear better clothes .... [On the contrary] those who stay behind will always remain poor". (NNJYZ 1995: 241–242).

Productivity

In addition to being a direct source of income, labor out-migration also has impacts on the productivity of rural labor. The labor that is lost, except for short periods of time when migrants return during planting and harvesting seasons, is the opportunity cost of migration. Presumably, the decision to migrate would not be made unless the return from migrant work is greater than the return from activities at the origin. A common way to evaluate this is to compare migrant income with non-migrant income. Using the 1995 CASS survey of rural household income (see note 51), Li (1999) estimates a household income function that controls for the effects of land, production assets, and location (province). The results show that the marginal contribution of migrant workers to total household income is higher, by about 10 percentage points, than that of non-migrant workers. Based on an average rural household income of 6,270 yuan in 1995, the estimate implies that an average migrant worker earns about 600 yuan more than household members that work at the origin. Also addressing the issue of opportunity cost, Hare and Zhao's (2000) survey of a rural county in Henan in 1995 finds that the marginal return to labor input to migration is higher than that to agriculture.[53] Using survey data from six provinces (Guangdong, Zhejiang, Hunan, Jilin, Sichuan, and Gansu) for the period 1986 to 1999, Yang's (2003) econometric modeling finds that the marginal product of grain production is only one yuan per labor-day, versus 17 yuan for migrant work in urban areas.

Not only does migrant work have higher productivity than non-migrant work, the former appears to have positive impacts on the productivity of those who remain in rural areas. Again using the income function, Li (1999) finds that the marginal product of non-migrants in migrant households is higher than that in non-migrant households. This finding suggests that migration induces reallocation of production resources within rural households in ways that increase productivity. For example, the remaining household members may be motivated to increase their efficiency in farm work. In addition, when migrants opt to lease out their farmland to others, they are indirectly fostering the scale economies of the leasers (Qiu et al. 2004; see Chapter 1).

Agricultural production

How labor out-migration affects agriculture is a controversial issue. Croll and Huang's (1997) study of eight villages in Jiangsu, Anhui, Sichuan, and Gansu, conducted in 1994, reveals negative effects of migration on agricultural production. They find that migrant work is seen as a quick fix that reduces population pressure and increases income, without requiring investment or in-situ development in poor areas. As a result, they argue, farmers tend to use migrant work to substitute for, rather than supplement, agriculture. Observations of uncultivated land, crop abandonment, and reluctance to introduce new crops, support the notion that farmers want to leave agriculture and have no desire to increase agricultural output. Furthermore, Croll and Huang warn that the widespread trends in agriculture substitution are likely to have serious repercussions for food sufficiency. Other researchers also link migration to land abandonment, reduction of agricultural output, and a dependent model of development (Ma et al. 2004).

The observation that migration has adversely affected agriculture is, however, not widely shared (Lian 2002). Based on econometric modeling and time-series data, Yang (2003) estimates that the impact of migration on grain production is less than 2 percent of the original grain output. This finding suggests that household members who remain respond to the loss of labor by working longer hours. Zhong (2000) observes that labor out-migration stimulates new and efficient ways to farm, and that migrants' remittances boost investment in agriculture. Along the same vein, Ma et al. (2004) document that grain production remained stable throughout the 1990s and suggest that this is due to improved productivity and increased capital input via remittances. All of this supports an observation made earlier, that migration improves the labor productivity of those who stay behind.

Agricultural labor

It is well documented that agricultural labor surplus in China is large. Estimates of the size of agricultural labor surplus are generally in the range of 150 million to 200 million and are expected to increase (Cai 2001; Cao 2001; Meng 2000; Qui 2001; Wang 2001; Zhang and Lin 2000).[54] Cai (2000: 187) estimates that about one-third of China's agricultural labor is surplus labor. Chen (2001) defines agricultural labor surplus as the difference between labor demand and supply in agriculture, or more specifically, the amount of labor supply above the point where marginal labor productivity is zero. Using arable land to evaluate labor demand, he estimates that, between 1962 and 1997, agricultural labor surplus increased from 42 million to 179 million. Chen notes, furthermore, that in 1997 labor surplus accounted for 52.6 percent of the total agricultural labor and that a full transfer of the surplus labor to non-agricultural sectors would have resulted in an output gain of 2.4 trillion yuan, an amount equivalent to 32 percent of China's GDP that year. Zhao (1998) estimates that the annual increase of rural labor force between 1997 and 2000 would be 6.6 million, of which one-sixth is surplus labor.

Labor out-migration is seen as a means to effectively reduce agricultural labor surplus (Wan 2001). Most researchers conclude that the impacts of labor loss from out-migration on agricultural production are insignificant. The dominant view is, therefore, that reducing the size of the rural population is key to resolving the *sannong* problem – *nongcun* (rural villages), *nongmin* (peasants), and *nongye* (agriculture) – the three aspects of rural China that are seriously lagging and underprivileged compared to urban China (Chen 2001; Han *et al.* 2002; Wang 2003; Wu and Cao 2002; Zhong 2000).

Yet, migration is selective. It is widely documented that the elderly, married women, and children are the ones that remain in rural areas, partly a result of the split-household strategy (see Chapter 5; Cao 1995; Cook and Maurer-Fazio 1999; Croll and Huang 1997; Davin 1998). Based on county-level data for 2000, Ma *et al.* (2004) show that as the rate of temporary migration increases from 2 percent to 15 percent, the proportion of persons aged 60 or older in the origin increases from 4.7 percent to 7.5 percent. Despite the out-migration of young labor, the general view is that this brain drain is not problematic because the remaining labor is more than enough to sustain agricultural production (Cai 2000; Lian 2002). From a somewhat different perspective, Ma (2001) observes that the human capital impacts of migration on the rural origin are stage-dependent. He argues that a brain drain at the time of migration will give way to a brain gain during migrant work and finally to a brain-drain reversal after migrants return.

Although the notion that labor out-migration adversely affects agriculture is generally not substantiated, there is evidence that the attractiveness of agriculture has declined. In Croll and Huang's (1997) study, farmers conclude that agriculture is an unprofitable activity and is redundant when the out-migration option is available. Due to the marked preference for non-agricultural activities and a strong desire to leave agriculture, Croll and Huang argue, out-migration would take place regardless of whether labor surplus exists. Indeed, other studies find that many rural–urban migrants have never engaged in agriculture and thus labor surplus is not a compelling reason for out-migration (*Guangdong wailai* 1995). Zhao (1998) observes that out-migration has become not only a strategy to raise income but a culture in some parts of western and central China, such that it is a popular source of "employment" for secondary school graduates.

Impacts of migration on rural areas 123

Zhong (2000) suggests that the opportunity for migrant work reduces peasants' motivation to invest in agricultural land, even though they continue to hold on to the farmland allocated to them. He refers to them as a "hybrid" class that straddles peasantry and urban opportunities, which echoes the notion that rural–urban labor migrants (and their households) seek the best of both worlds – urban work to boost income and improve rural living, and the countryside for insurance, security, and long-term settlement (see Chapters 1, 5 and 9).

Poverty and inequality

Labor out-migration is widely believed to be conducive to alleviating rural poverty. Estimates of rural population in poverty in the mid-1990s were in the range of 70 to 80 million (Cai 2001: 327; Fan 1997), and the Chinese government has indeed identified rural poverty as an urgent problem (see Chapter 9). Numerous studies have found that migrant households' income is higher than that of other rural households.[55] Fan, L. (1997) provides a convincing analysis of the relationship between migration and poverty alleviation. He argues that out-migration from poor areas, where the carrying capacity of land is small, is necessary for relieving population pressure. In addition, Fan observes, migration improves the quality of rural labor and is therefore a more effective means of poverty alleviation than direct capital transfer.

Evidence on whether labor out-migration can reduce rural–urban and regional inequalities is, however, inconclusive. On the one hand, migration raises income and productivity in less-developed regions and is thus expected to alleviate regional inequality (Ma *et al.* 2004; Rozelle *et al.* 1999). Furthermore, the flows of skills, knowledge, and ideas via migration to the origin are conducive to its development (Lian 2002). On the other hand, skeptics point to effects of migration that may exacerbate regional inequality. Lian (2002) argues that migration replenishes labor, keeps wages low, and increases market demand in the more developed coastal areas and that it discourages profitable labor-intensive sectors from moving inland. He predicts that migration will enlarge regional disparity. Using a spatial model, Hu (2002) shows that migration accelerates industrial agglomeration in coastal areas and in turn widens the coast–inland gap. Finally, Hare and Zhao (2000) find that the return from rural labor allocated to migration is higher than that in agriculture but lower than that in local non-agricultural activities. Thus, they argue that migration alone is not enough to stimulate local economic growth and diversification and they caution against seeing migration as a substitute for locally based efforts to improve economic conditions in rural areas. In the same vein, Croll and Huang (1997) observe that remittances are inconsistent and do not necessarily contribute to the development of the origin village.

In summary, existing evidence strongly supports the notion that labor migration has raised rural income, improved rural productivity, and alleviated rural poverty. The impacts of migration on regional and rural–urban inequalities and agricultural production are, however, inconclusive. It is clear that migration is an attractive and profitable alternative to agriculture. Whether labor out-migration has long-lasting negative impacts on agriculture remains an open question. One of the factors is whether peasant migrants will stay in the destination or return. The next section focuses on return migration.

Return migration

Since the late 1990s, researchers have begun to pay greater attention to the impacts of return migrants on rural areas. Estimates of return migration vary but are generally in the range of 20 percent to 40 percent. Based on analysis of data from the 1995 One Percent Population Sample Survey, Liang and Wu (2003) estimate that about 20 percent of all the interprovincial migrants originating from Sichuan have returned and that about one-third of all the migrants from Sichuan to Guangdong have returned. Cai (2000: 204) notes that about one-fourth of out-migrants from northern Jiangsu have returned. Studies on Shandong and Anhui document that by the late 1990s about 20 percent of rural migrants had returned (Bai and Song 2002: 8). The rate of return migration, based on the 1999 Sichuan and Anhui Household Survey, is 28.5 percent (Bai and Song 2002: 15, 27). Murphy (2002: 125) cites a study that reports that 36 percent of rural migrants from the inland provinces of Jiangxi, Anhui, Hubei, and Sichuan have returned. An even higher percentage – 38.4 percent – of return migration is reported by a rural household survey in six provinces (Hebei, Shaanxi, Anhui, Hunan, Sichuan, and Zhejiang) conducted by the Ministry of Agriculture in 1999 (Zhao 2002).

Estimates of return migration are difficult to compare, however, as the definition of return migration varies considerably from one study to another. The exact time of initial migration and the duration at the destination are often not known (Liang and Wu 2003). Different criteria have been used to distinguish return migrants from circular migrants. A common assumption is that the longer migrants have returned and stayed at the origin, the less likely it is that they will go out again. Zhao (2002) defines returnees as migrants who have returned and stayed in the place of origin for at least eight months. Based on the same data source, Bai and Song (2002: 11–15) use nine months and Wang and Fan (2006) use 12 months as the minimum requirement for defining an individual as return migrant as opposed to circular migrant. Yet another variation is definition of the place of origin. While some studies, especially those conducted at the origin, focus on returnees to the origin village (e.g. Bai and Song 2002: 11), others define all those who return to the origin province – not necessarily to the origin village – as return migrants (e.g. Liang and Wu 2003). Despite these complexities and variations in definition, most studies do depict that a significant proportion of rural–urban labor migrants have returned.

Evidence on the impacts of return migration is mixed. No consensus has been reached on whether returnees are primarily successes or failures and the extent of their impacts on the rural economy. Success returnees are generally

Impacts of migration on rural areas 125

understood as those who have thrived in the destination and who choose to return to the origin. Failure returnees, on the other hand, are those who have not achieved their goals, are rejected by the destination or are forced to return (Gmelch 1980; King 1986; Lee 1984). Studies of rural–urban migration in other countries have found that migrants who have difficulties finding jobs in cities and adapting to city life are more likely to return (Borjas 1989; Newbold 2001; Reyes 1997; Simons and Cardona 1972: 168). In other words, returnees tend to be failures rather than successes. Most studies on China, however, challenge this view. Cai (2000) argues that return migrants are not failures because many rural–urban migrants do not intend to stay in the destination for good in the first place (see also Chapters 5 and 6). Based on a 1995 survey of 309 households in Henan, Hare (1999a) concludes that failure at the destination is not an important reason for return.

In addition, researchers have examined the role of returnees as entrepreneurs and in rural economic activities. Based on a survey of 119 villages from 13 counties across nine provinces, conducted by the Department of Rural Development of the Development Research Centre of the State Council and including interviews with more than 2,000 returnees, Ma (2001) argues that return migration represents a brain-drain reversal to the rural origin. He underscores the leading role returnees play in transforming their rural communities, by diversifying the economy and promoting entrepreneurial activities. Furthermore, Ma shows that migrants who value skill improvement more than income gain are more likely to engage in non-farm work upon their return, supporting the notion that the human capital improvement of returnees has long-lasting impacts on rural development. Based on the same data source, Wang *et al.* (2003) observe that most returnee entrepreneurs are relatively young and well educated and that return migration is an important factor in increasing employment opportunities in rural areas. The notion that migrants return to invest has become increasingly popular (Zhao 1998). Zhao (2002) notes that return migrants are more likely than non-migrants to invest in farm machinery. Other studies find that returnee entrepreneurs invest in a variety of manufacturing and services sectors as well as agricultural businesses (Bai and Song 2002: 129). Murphy (1999; 2000) emphasizes returnees' roles as agents of information transfer, entrepreneurship, and economic diversification in their natal communities. She finds that return migrants are effective in negotiating with local governments, which in turn would improve the local business environment.

Exposure to new ideas and technology in urban areas has indeed enabled some returnees to start innovative projects in the home community. This 35-year-old Sichuan woman's narrative describes one of these success stories:

> My father took me and my sister to Xuzhou, Jiangsu to sell cooked ducks .... I observed that urban people like to drink milk. I also learned about the health benefit of milk ... . There were no fresh milk factories in my home area Caohu, so I decided to return and set up a fresh milk factory .... I invested a

large sum of money in advertising the benefit of milk. Many people here have since accepted the concept and are now drinking milk every day.

NNJYZ 1999: 557–559

Another 31-year-old woman also started an innovative activity in her home community:

> I went to Shanghai and worked as a tailor for a garment factory. I observed the many efforts urban parents put into their children's education as well as the high expectations they have for the children .... All urban children receive preschool education .... I decided to open a preschool in my home area .... The school now serves the need of 20 children from three nearby villages.
>
> NNJYZ 1999: 192–193

This man's account shows that skills acquired from migrant work can result in technology transfer to the countryside:

> I migrated to Baoshan District in Shanghai ... and learned almost all the technology in vegetable planting .... After I returned to my village, I decided to grow greenhouse vegetables .... Since then my family income has increased dramatically. Farmers in my village then learned this technology from me ... [and] as a result their income has increased greatly.
>
> NNJYZ 1999: 532

Despite the general optimism in the literature and success stories such as those described above, there is little systematic evidence to support the assumption that success returnees are highly represented among return migrants. Bai and Song (2002: 42, 129), for example, document that entrepreneurs are extremely rare among return migrants in Sichuan and Anhui. In the 1999 Sichuan and Anhui Household Survey, only 2.5 percent of return migrants identify investment as one of the reasons for return. Based on the same data source, Wang and Fan (2006) find that returnees are older, less educated, and less likely to have received training for non-agricultural work than continuing migrants. In the same vein, Liang and Wu (2003) show that return migrants in Sichuan are negatively selected and that there are no significant differences between returnees and non-migrants in non-farm employment. Similarly, Zhao (2002) concludes that return migrants are no more likely than non-migrants to engage in non-farm activities. Despite Murphy's (1999; 2000) observation that returnees collaborate with the local state in rural development, she cautions that their role in creating jobs for the non-migrant population and in non-agricultural activities is limited.

The small but growing body of literature on return migration, therefore, has not produced conclusive findings on returnees' impacts on the countryside. This in part reflects a relative lack of attention to the reasons for return migration. Based on the 1999 Sichuan and Anhui Household Survey, Wang and Fan (2006) find that the most prominent reasons for migrants to return are difficulties finding jobs or making money and the need to care for family members. Both reasons highlight the institutional and social exclusion that migrants confront in cities. This 40-year-old Sichuan woman, who worked as a street-sweeper in Beijing, describes why she returned: "My monthly wage was only 300 yuan .... The money I earned was not enough to help me make ends meet .... It is very difficult to make money in Beijing if one does not have connections" (NNJYZ 1999: 139–140).

Children's education is a common reason for return, as schools in cities charge migrant children higher fees than local children or simply do not accept migrant children. Education for migrant children, who amounted to more than 14 million in 2000 and almost 20 million by 2005, is rapidly becoming a burning question in Chinese cities (Wang, Fang 2005). Both national-level and case studies have found that children in the floating population are much less likely to be enrolled in school compared to non-migrant children (Liang and Chen 2007; *Zhongguo renmin daxue* 2005). Migrant children are required to pay higher fees in cities even though they are from the same province, as described by this 40-year-old Sichuan man: "It is much more expensive for non-Chengdu children to go to school there .... We could not afford it and had to return so that our children can continue their education" (NNJYZ 1999: 257).

An alternative for peasant migrant families is to send their children to "migrant children schools" usually organized by migrants themselves (Kwong 2004). It is estimated that in 2004 there were 280 such schools in Beijing alone, enrolling about 50,000 migrant children – representing only one-fifth of all school-aged migrant children in Beijing (Li 2004). The vast majority of these schools are not licensed and their quality is low. Sending the children to schools in the home village or town is considered by many migrants a better option, as described by this 33-year-old Sichuan woman:

> My husband and I had worked in Shanghai for six years. When my children reached school age, we returned home .... I was not satisfied with the quality of the teachers at the Shanghai Dagong [Migrant Children] School .... We returned for the sake of the children's education.
>
> NNJYZ 1999: 53

The needs of family members are, therefore, central for understanding decision-making behind return migration (see also Chapter 5). Wang and Fan (2006) argue that the conventional success–failure framework for explaining return migration is inadequate and must be supplemented by a family perspective. Often, multiple reasons underlie migrants' decisions to return, reflecting their balancing responsibilities and resources in the home village, as illustrated by this 37-year-old Sichuan man (see also Bai and Song 2002: 43):

> My wife stayed home to farm and my son goes to school at the village .... I returned from Shanghai because first, both of my brothers went out to work and nobody was taking care of their farmland, and second, my parents are elderly and need someone to take care of them.
>
> NNJYZ 1999: 92

This example shows once again that mobility decisions involve negotiation and collaboration among family members, including members of the extended family and those not living under the same roof (see Chapters 1 and 5).

Social impacts

Migrant work represents more than an important opportunity for economic gains. Peasant migrants not only earn money for themselves and the family but also are exposed to new ideas and perspectives. Although there is little systematic evidence on the social impacts of migrant work, migrants' narratives show that multifaceted social changes are taking place in the countryside as a result.

Migrant work and independence

The traditional rural production system that centers on agriculture offers few opportunities for social and economic mobility. To peasant women, in particular, whose lives revolve around care-giving, house chores, and farming, the contributions of their labor are often assumed but not recognized. Off-farm wage work, on the other hand, represents a new reward system and enables individuals to make significant, remunerated financial contribution to the family. To young peasant women who are expected to be subordinate to their parents until marriage and then to their husbands upon marriage (see Chapter 1), migrant work and experience enable them to gain exposure and above all a sense of achievement and independence. This 18-year-old Sichuan woman who works in an eye-glasses factory in Shenzhen comments: "By working as a migrant I can support myself, rather than depending on my parents" (NNJYZ 1995: 59–61). This 21-year-old Sichuan woman relates migrant work to independent thinking: "I found the experience rewarding. I earned my own living. I was naive when I was staying in the village. When I was away from home I had to be independent and make my own decisions" (NNJYZ 1995: 174–176).

The ability to support themselves and to work and live independently from the parents can also improve women's status and their self-image, even after they have returned from migrant work, as illustrated by this 23-year-old woman's narrative:

> After I returned, I felt that my parents respect me more since I had lived outside [the village] independently for a long time. They now allow me to participate when they discuss family affairs .... My migration experience

has improved my working and living abilities. I feel that I can handle things better than before and I can get along with people better as well.

NNJYZ 1999: 396–397

Broadening one's exposure is an objective and outcome mentioned by some peasant migrants (see also Table 4.8). For example, this 22-year-old Sichuan woman remarks:

Young people should try to go somewhere and increase their exposure .... Women, especially, should try to do migrant work. If they stay home there's no work for them other than house chores, farm work and going to the market. When you do migrant work you have income and can help the family.

NNJYZ 1995: 176–180

Similarly, this 22-year-old woman portrays migrant work as seeing the world and having fun: "I migrated to Shanghai with my husband after we got married .... I had a lot of fun in the outside world and did not miss home" (NNJYZ 1999: 93–94).

It appears that peasant women, more than their male counterparts, make the connection between migrant work on one hand and independence and exposure on the other. This is due to deep-rooted patriarchal ideology that determines gender differentials in opportunity in both economic and social spheres: peasant men have traditionally had more opportunities to engage in off-farm and wage work than women; men are expected to be the breadwinner and women are dependent on them; and men are assumed to be responsible for the outside and women for the inside (see Chapter 1 and "Gender division of labor within marriage"). Migrant work offers opportunities to challenge the above traditions and can be an empowering experience for peasant women. Recent studies have, likewise, stressed the importance of migrant work in increasing young, rural women's sense of independence and their economic value to parents (Jacka 2006: 134; Zhang 2007).

Gender division of labor within marriage

As discussed in Chapter 5, the migrant work duration of single, rural women tends to be short because they marry young and most return to the countryside to get married. Upon marriage, they are expected to stay in the village to farm and to raise children. Their husbands, on the other hand, can continue to pursue migrant work. Marriage, therefore, signals a termination of migrant and "outside world" experience for rural women, as illustrated by this 22-year-old woman's remark:

Right now, my child is too young and I have to stay home .... What I want to do most is to go out and work. I do not want to stay home taking care of the child, but I have no choice.

NNJYZ 1999: 93–94

130 *Impacts of migration on rural areas*

The demographics of villages that have large numbers of out-migrants have accordingly changed, as described by this Sichuan woman (Plate 7.1): "Most young people in our village have migrated to work. Some middle-aged persons have done so as well. Those who stay behind are mostly the elderly and middle-aged women" (NNJYZ 1995: 94–96).

The split-household strategy, therefore, reinforces gender division of labor and reproduces the age-old inside–outside ideology in a modified version – the inside includes not only the home but also the countryside and agricultural work, and the outside refers not only to what is beyond the home but also to urban, non-farm opportunities and experience (see also Chapter 5). Whereas migrant work may be empowering to young, rural women, marriage aborts this path of social and economic mobility and relegates them back to the village. On the one hand, the split-household arrangement gives peasant women opportunities to make farming and other decisions on their own (Davin 1998). On the other hand, they are left with multiple and increased workloads, as illustrated by this 23-year-old Sichuan woman: "Many men from our village are migrant workers. The women and elderly in their households sometimes have difficulties dealing with heavy work" (NNJYZ 1995: 26–28).

The split-household strategy may present a dilemma to wives of migrant workers – they welcome the remittances but miss their husbands and are overwhelmed by multiple responsibilities. This woman from my Gaozhou field study

Plate 7.1 In some Chinese villages, most young men and single women have left for migrant work, leaving behind married women, their children, and the elderly (source: Author, Gaozhou, January 1999).

(see Chapters 1 and 8) was able to build a new house with remittances from her husband but she is overworked and rarely sees her husband:

> My husband rarely returns home – maybe two or three times a year. Because he is away, no one is helping me. I have to work all day long. I am always busy, taking care of house chores, the children, and the fruit trees. I contact my husband by phone about once a month, but he lives in a factory dorm and is hard to reach. It would be nice if he stayed home more. I've asked him to return home for good, but then his work is bringing home money.
> Author's interview, Gaozhou, Guangdong, December 1999

Furthermore, agricultural work is increasingly considered inferior to non-agricultural activities. Both local communities and rural households may be reluctant to invest in agricultural infrastructure and labor, thus further intensifying the vicious cycle of de-skilling of peasant women and disdain for agriculture (Croll and Huang 1997). All of this suggests that the spatial and occupational division of labor between peasant women and migrant husbands would further undermine the status of women in the countryside.

Marriage and marital disruption

The outside world exposes rural men and women to ideas that challenge age-old concepts about marriage. While the norm is still marrying someone from the countryside and not too distant from the home village (see also Chapter 8), migrant work has motivated some rural Chinese to consider finding a marriage partner elsewhere, perhaps as a means to leaving agriculture and poverty. This 23-year-old Anhui woman comments:

> I don't want to marry a farmer. Farming is meaningless. You work from the beginning to the end of year and you still don't have enough money to spend. Farm work is heavy … dirty and exhausting. I want to marry someone outside [my native place], but I have not yet met a suitable person.
> NNJYZ 1995: 509–511

Her strategy – using marriage as a means to achieving physical and economic mobility – whether successful or not, is not uncommon among peasant women. In Chapter 8, I shall examine in detail how peasant women pursue marriage migration to improve their well-being.

Observations about urban lives may lead peasant migrants to reevaluate gender relations within marriage. Direct challenges to traditional gender roles, such as reverse division of labor, where the wife does migrant work and the husband stays in the village, are controversial and hotly contested (see Chapter 5). However, exposure to urban areas appears to have brought about subtle and gradual changes in views toward gender roles, as illustrated by this 31-year-old man:

> During my migrant work, I noticed that husbands in cities cook and do laundry. That opened my eyes and so I began to help my wife with housework. Before that, I had never done any housework. Now, I work two to three hours a day on house chores. Both my wife and I have control over family income and savings.
>
> NNJYZ 1999: 312–314

Similarly, this 49-year-old man speaks about his taking up more housework:

> When I was in Beijing, I was very impressed with the husbands there, since they took care of some household chores. So, after I returned I began to help my wife with housework. I talk to my wife a lot and discuss family issues with her.
>
> NNJYZ 1999: 135–137

This 32-year-old woman describes how the way women are treated differs between urban and rural areas and evaluates which is better:

> Beijing people like to stand up to protect women. If a man has a fight with a woman on the street, people will blame the man and try to stop him. If this happens in the countryside, people will just watch and do nothing … . Spousal relations are different between urban and rural families. In rural areas, if a husband beats his wife or quarrels with her, it's nothing. But an urban woman would not let it go. She may even divorce the husband. From this point of view, rural areas are better than cities since we have a lower divorce rate.
>
> NNJYZ 1999: 223

Although she points out that urban women are treated with greater respect compared to their rural counterparts, her comment on divorce reflects a common perception about the city – that it is disorderly, corrupt, and fraught with bad influences (Gaetano 2004; see Chapter 5). This 30-year-old man elaborates these perceptions:

> There are negative impacts of migration … . Women going out is bad for family relationship. After staying in cities for an extended period of time, women expect more [from their husbands]. They are not satisfied with the husband back at the village any more. There are cases in this village that such women ran away with other men.
>
> NNJYZ 1999: 81–83

Another 60-year-old man contrasts migrant and non-migrant families: "Spousal relationship differs between migrant and non-migrant families. In this village, there are no divorces in non-migration families. All divorces occur in families that have had migration experience" (NNJYZ 1999: 356–358).

The above comments, while evaluative of migration's impact on marital stability, reveal a deep commitment to the institution of marriage, one that is central to the social fabric of the countryside. This commitment appears to be intact, despite the rise in divorce rate in recent decades, especially in large cities. However, there is little systematic evidence about how the geographical separation between husbands and wives – a centerpiece of the split-household strategy – affects rural marriages. This is partly because surveys in the countryside, which is much more conservative than the city, are rarely able to examine directly personal and sensitive topics such as marital disruption. My field study in Gaozhou (see Chapters 1 and 8) shows that some male migrants, especially financially successful ones, have extramarital affairs in the city. The wives that stay behind in the village may suspect or know about the affairs but most choose not to raise issues with their husbands. When asked why they tolerate such situations, a typical response is because the husband continues to send home remittances, as summarized by one of my informants:

> They want their husbands to stay in the village, but they'll remain poor if the husbands return. The wives have to stay in the village because of the children. Some men don't want to return. Most women's mentality is, "so long as the husband takes care of the home and the children [by sending money] I am contented."
>
> Author's interview, Gaozhou, Guangdong, December 1999

These observations, though anecdotal, support not only the notion that rural Chinese are committed to protecting the marriage institution, but also the argument made earlier that the split-household strategy reinforces peasant women's marginalization imposed by their lack of economic independence.

New ideas and lifestyle

Studies have shown that peasant migrants have not only improved their material well-being but are also fostering social change in the countryside, by importing new lifestyle and value from urban areas, especially from places of migrant work (e.g. Cao 1995). This 22-year-old Anhui woman who works in domestic work in Beijing comments:

> Beijing is wonderful. It is big, sophisticated, and it has so many things not found in the village … . Through migrant work we learn skills and knowledge and make money … . Our quality of life has improved as well. In particular, we dress much better now. Migrant work increases our exposure, so that we talk more politely and we become wiser. Villagers imitate migrants. They buy western-style furniture, and decorate their homes following the styles in Beijing … . Villagers now pay more attention to sanitation. Their taste for food has also changed. (see also "Return migration")
>
> NNJYZ 1995: 339–341

Another 46-year-old Anhui woman has also observed changes in clothing and sanitation: "Migrant workers have learned a lot. Peasants now are more knowledgeable about sanitation and more particular about clothing. Some people now change their clothes more frequently"[56] (NNJYZ 1995: 352–354).

In villages that have benefited from labor out-migration for an extended period of time, there also are changes in social practices such as weddings, funerals, and dialect, as illustrated by this 42-year-old Sichuan woman:

> As more people are going out to do migrant work ... weddings and funerals have become more elaborate and classy. Young migrants' conversations and manners have become more civilized; they can now speak *putonghua* [Mandarin], not like us whose dialect people don't understand.[57]
>
> 2005 Sichuan and Anhui Interview Records

Migrants' outlook may also change as a result of urban experience. This 23-year-old Anhui woman describes some of these changes:

> Many changes have occurred to migrants. They turn huts into large houses, dress better and are more open. Those who stay behind look less smart. As far as meeting and seeing people of the opposite sex, migrants are more open, but non-migrant women tend to remain feudalistic [*fengjian*]. Many migrants have learned some skills and figured out a way to make money, and they build houses and buy electrical appliances. They eat better now, and kill a pig [to eat] during the Spring Festival. Villagers used to be simple-minded, but are now more cunning .... Older cadres in the village are too slow and conservative. They should retire and let young people do their job.
>
> NNJYZ 1995: 294–296

Her critique of village cadres is an example of migrants reflecting on and evaluating rural life because of their increased exposure. Furthermore, migrants may perceive non-migrants as unsophisticated. This 23-year-old Anhui woman, for example, feels distant from non-migrant women:

> When I return to the village I usually hang out with other young women migrants. I have little to talk about with non-migrant women. We have different lifestyles, and we don't see eye to eye. To us [migrants], non-migrant women wear reds and greens[58] and look unrefined. To them, our permed hair makes us look strange and alien.
>
> NNJYZ 1995: 509–511

Similarly, this 22-year-old Anhui woman is critical of other villagers:

> I don't respect those who prefer staying home doing nothing to migrant work. More than 90 percent of the young women in our village do not go to

school. When they go to Beijing they look especially dumb and are likely to get into trouble.

NNJYZ 1995: 339–341

Critical evaluations such as the above may prompt migrants and returnees to bring about changes to the family and to the village. This 22-year-old Sichuan woman's comment supports the notion that migrants are agents of social change in the countryside: "After migrants have lived in more developed areas, their attitudes change. They become dissatisfied with their situation and want to improve their family's well-being by working harder" (NNJYZ 1995: 176–180).

Summary and conclusion

Labor out-migration and return migration have potential to bring about profound changes in the Chinese countryside. Migration's existing and future impacts warrant greater attention by researchers, as the Chinese population is still primarily rural, many rural areas are persistently poor, and rural–urban inequality is large.

This chapter has reviewed the economic impacts of migration in three areas – rural income and productivity, agricultural production and labor, and return migration. There is overwhelming evidence that labor out-migration raises rural income and productivity. Remittances, in particular, have become an important means to improve economic well-being, to finance the economic and social activities of peasant households, and to alleviate rural poverty. It is unclear if migration reduces or widens rural–urban and regional gaps, however, as it contributes to both rural income growth as well as economic growth in cities and coastal areas. Second, the impacts of labor out-migration on agriculture are inconclusive. While farmers may substitute migration for agriculture, the large rural labor surplus and improvement in agricultural productivity suggest that adverse effects on agricultural production are minimal. Third, the small and emerging body of work on return migration shows that significant proportions of rural–urban migrants have returned but their impacts are mixed. While some returnees have transformed themselves into entrepreneurs, many have returned because of institutional and social exclusion by the city. Migrant narratives suggest that the success–failure dichotomy for interpreting return migration is too simplistic and must be supplemented by a household perspective. Mobility decisions, including return migration, are heavily driven by the needs of the household as a whole and those of specific family members.

Social impacts of migration are equally multifaceted and mixed. Migrant work enables peasants, especially young, single women, to gain a sense of self-worth and independence. However, the split-household strategy reinforces the inside–outside dichotomy within marriage and binds married women to non-wage and unskilled work in the countryside. Urban views may

136 Impacts of migration on rural areas

challenge traditional gender roles and relations, but these are often seen as potentially disruptive of the deep-rooted social order in the countryside. Finally, migrants are bringing back urban lifestyle and value to their villages, which can be bases not only for improvement but also for critique of rural lives.

8 Marriage and marriage migration

Introduction

Most studies on migration focus on labor and work-related moves. Relatively little attention has been given to marriage migration (Watts 1983). Typically, marriage is treated as one of the life events that trigger migration, and marital status is considered one of many factors that explain mobility differentials (Mulder and Wagner 1993; Speare and Goldscheider 1987). Marriage migration, defined as migration to join the spouse in another area usually at or soon after marriage,[59] is often considered a special kind of migration for the goal of family formation (e.g. Lievens 1999). This chapter examines the relationship between marriage and migration in China and illustrates in particular how marriage migration sheds light on gender roles and relations in the household on one hand and women's agency on the other.

Marriage and migration

A dominant view about the relationship between marriage and migration is that women are tied movers (passive movers or trailing spouses), who migrate as a result of marriage or within marriage (Bonney and Love 1991; Fincher 1993; Houstourn et al. 1984; Oberai and Singh 1983; Thadani and Todaro 1984; Watts 1983). Indeed, the vast majority of marriage migrants are women (UN Secretariat 1993). The bargaining-power hypothesis postulates that it is more often the woman than the man who bridges the distance, because the husband's occupation is considered more important (Mulder and Wagner 1993). Patriarchy within marriage and in the larger society, manifested as the power difference between men and women within marriage, the shorter-term career prospect of women due to their expected care-giving roles, and the persistent gender wage gap in the labor market, is central to explaining women's greater likelihood to be tied movers.

On the other hand, patriarchy also underlies the notion that women may enter into marriage in order to pursue a better chance of economic security and well-being (Walby 1990). Put in another way, given their marginalization in the labor market, women may find marriage, and specifically hypergamy (marrying up), an attractive and perhaps even the only vehicle toward economic betterment.

In this light, marriage can be regarded as not an end in itself but a means to attaining certain goals (Willekens 1987). When these goals involve moving to more desirable locations, marriage can be an instrument for, rather than an objective of, migration.

Some studies have examined marriage as an opportunity for migration and related to that the role of location in marriage migration (Humbeck 1996; Rosenzweig and Stark 1989; Thadani and Todaro 1984). Research on international marriage migration, in particular, has focused on the destination's advantages to marriage migrants. Humbeck (1996), for example, finds that Thai women who migrated to marry German men did so in order to escape poverty in rural Thailand, help their natal families, and seek better employment opportunities. Piper (1999) shows that getting secure and legal residence status, including status for work, in the host country is an important factor of international marriages of women from Korea, Thailand, and the Philippines to Japanese men. Studies of internal migration also have linked marriage migration to economic opportunities. Riley and Gardner (1993) observe that, in many societies, marriage is a strategy for the wife's natal family to improve its status and that the location of prospective sons-in-law is important. Rosenzweig and Stark's (1989) study shows that households in rural India favor marriage of daughters to distant and kinship-related households in order to mitigate income risks. Likewise, Watts (1983) highlights the desire for rural women in Nigeria to marry men in cities.

The notion that marriage is a strategy stresses the agency of women. Focusing on Puerto Rican women, Ortiz (1996) argues that they are active agents in the migration process, using it as an economic option and to gain independence. Lievens's (1999) study of Turkish and Moroccan women in Belgium who marry men from their countries of origin suggests that they do so in order to secure more independence and to free themselves from the influence of in-laws.

The concept that marriage is a means toward an end, including migration, highlights the marriage decision-making process as one involving pragmatic considerations and weighing of options, complicating the conventional view that emphasizes marriage as a matrimonial union based on affection. Pragmatism also underlies the decision-making of those who marry migrants. Humbeck (1996), for example, observes that German men who married women from Thailand are older and less well educated, suggesting that their limited marriage market induced them to seek foreign wives. In rural Japan, demographic changes have contributed to a "wife shortage" and a marriage squeeze among men and encouraged the importation of wives from Southeast Asia (Jolivet 1997: 152–161; Piper 1999). Although the specific contexts vary, a common thread in these studies is that men that have difficulties finding mates locally are more likely than other men to marry migrant women.

The literature on marriage and migration, though growing, is small and has two major gaps. First, much of the attention is paid to international marriages. There are relatively few empirical studies on marriage migration within countries. Second, there is very little attention to socialist and transitional economies. Thus, how the institutional contexts of, and structural changes in, these economies affect marriage decision-making and marriage migration is not well understood.

Marriage and migration in China

As described in Chapters 1 and 5, age-old traditions that undermine women's status remain strong in China, especially in the countryside. Large gender gaps in education, the constraints women face in the labor market, and the strong connection between marriage and women's well-being persist (Bauer *et al.* 1992; Bossen 1994; Honig and Hershatter 1988; Lavely 1991; Wang 2000; Wang and Hu 1996: 287). The Confucian ideology that governs gender roles stresses the importance of marriage for women. The notion that a woman's *xingfu* (well-being or happiness) depends on her marriage has been one of the cornerstones of gender roles and relations in China. In addition, marriage has always been considered an inevitable and indispensable life event, hence the old saying "when boys and girls reach adulthood they should get married" *(nanda danghun nuda dangjia)*. These traditions are especially steadfast in the countryside. The patrilocal tradition and the transfer of the wife's membership through marriage further reinforce the centrality of marriage to a woman's well being. Most peasant women have few alternatives other than marriage in order to escape poverty and achieve upward mobility (Bossen 1994; Honig and Hershatter 1988; Wang 2000; Wang and Hu 1996: 287). Indeed, historically, hypergamy has been common among women in China (Croll 1981: 97).

For thousands of years, however, peasant women in China have not married far. The most common practice has been to marry men in nearby villages, which serves at least two purposes. First, it allows the natal family to enlarge its networks in relative proximity. Second, marrying someone from outside the home village – village exogamy – minimizes the possibility of kin marriages, since many villages in China are dominated by households sharing the same lineage (Davin 1999: 141–142; Potter and Potter 1990: 205). The small geographical size of peasant women's marriage market also reflects their low levels of spatial mobility and limited knowledge about other places. This form of traditional marriage migration occurs over short distances and remains prevalent in the countryside (Wang and Hu 1996: 283; Yang 1991; Zhang and Zhang 1996). A survey in the late 1980s found that most rural marriages did not exceed a 25 km radius *(Renmin ribao* 1989). Since villages in close proximity tend to have similar economic conditions, these types of short-distance marriages rarely bring about significant improvement in well-being for marriage migrants. In the past two decades or so, however, long-distance marriage migration has become more prevalent. I argue that such migration reflects changing circumstances due to economic reforms as well as the agency of women to advance their well-being through marrying "into" more favorable locations. But first let me turn to the pragmatic considerations associated with marriage.

The process that leads to marriage in rural China has been, and still is, pragmatic and transactional. Economic functions and family considerations have always defined marriage – for the continuation of the family line, increase in family labor resources, formation of networks, provision of old-age security, and transfer of economic resources (Croll 1984; Ebrey 1991; Potter and Potter 1990; Wolf 1972). Although the Marriage Law of 1950 abolished arranged marriages (see also Chapter 1), in rural China marriage continues to be considered a transaction between two families who are assumed to act rationally to maximize their advantages (Croll 1981; 1984; Lavely 1991). For example, bride-price connotes compensation from the husband's family to the natal family for raising the daughter and for giving up her membership and labor. Dowry, on the other hand, is a gift from the natal family to the husband's family (or the newly wed) for taking responsibility for the daughter's livelihood.

Pragmatism and transaction are operationalized through a mate-selection process, which involves, explicitly or implicitly, evaluation of a potential spouse's attributes (*tiaojian*), such as age, education, occupation, income and economic ability, physical characteristics (e.g. height, appearance), health, class, personality, family background, and resources (Fan 2000). Attributes may also be functions of specific political–economic contexts. During the Maoist collective period, for example, former landlords connoted bad class origins and an undesirable attribute for marriage, while Communist Party membership was considered a good attribute (Croll 1981: 86–93).

The evaluation process may include both attribute matching and attribute trade-off. Attribute matching centers on the relative similarity in status between the prospective spouses, subject to the hypergamous principle that husbands should be "superior" to wives in terms of age, height, education, and occupation[60] (Ji *et al.* 1985; Lavely 1991; Tan 1996b; Yang 1994: 220). However, it is also widely believed that marriages involving households with similar socioeconomic statuses (*mengdang hudui* or "matching doors") are more stable and successful than those between households of very different statuses.

Attribute trade-off refers to the ways in which a desirable attribute can compensate for or offset a less desirable attribute. For example, a prospective husband's older age may be offset by his wealth. Trade-off can also facilitate matching, if a desirable (or less desirable) attribute of a prospective spouse is offset by another desirable (or less desirable) attribute of the other spouse. For example, a prospective husband's lack of education may be offset by the prospective wife's less desirable physical appearance. The emphasis on attribute matching and trade-off partly justifies the use of a third party – an intermediary or a go-between. The intermediary is often called the "introducer" (*jieshao ren*) or matchmaker (*meiren*) (Croll 1981), whose role is to bring together single men and women at marriageable ages[61] (see also Chapter 5) and that are fitting (*dengdui*) to one another. Intermediaries can be neighbors, relatives, friends, *tongxiang* (see Chapter 6) or professional brokers and are usually women. The intermediary may evaluate *tiaojian* matching and trade-off before the prospective spouses meet, or he/she may initiate that process via a meeting (*xiangqin*) between the prospective spouses. Either way, the role of intermediaries is crucial. In cases where prospective spouses meet on their own, their parents may still insist that an intermediary be identified to formally begin the marriage process.

Location is one of the attributes for consideration in the mate-selection process. The condition of the prospective husband's village is an important factor in a peasant woman's marriage decision-making. In what follows, I show that changes in the reform period have further reinforced location as a marriage attribute, that long-distance marriage migration reflects both structural forces as well as peasant women's agency, and that labor migration has important implications for marriage and marriage migration in rural China.

Marriage migration in China

The magnitude of marriage migration in China alone is sufficient reason for researchers to take it seriously. Marriage is the leading reason for female migration according to the 1990 census, accounting for 4.4 million of female intercounty migrants and 1.4 million of female interprovincial migrants (see also Figure 5.2 and Table 8.1). By the 2000 census, the number of female interprovincial marriage migrants has increased further to 1.6 million.

While the traditional short-distance village-to-village marriages are still dominant in rural China, studies have shown that long-distance marriage migration up the spatial hierarchy is increasingly prominent (Davin 1999, p. 145; Fan and Huang 1998; Fan and Li 2002; Tan 1996b; Wang 1992; Yang 1991). Although spatial hypergamy – women moving to more prosperous areas through marriage – is not new in China (Lavely 1991), I argue that circumstances in the reform period have reinforced the role of location in the matching and trade-off considerations underlying marriage migration. Second, I argue that the surge in rural–urban labor migration complicates the relations between marriage and migration. Specifically, increased mobility through labor migration increases the size of the marriage market and is conducive to household division of labor within marriage. The following two sections articulate these arguments.

Spatial hypergamy and marriage migration

Spatial hypergamy has always existed in rural China (Lavely 1991). Men in prosperous villages are more appealing than those in poor villages, as women desire to move up the spatial hierarchy through marriage. Long-distance marriage migration, however, has not been common until recent decades. The increasing prominence of long-distance marriage migration is at least partly related to recent structural changes. Since the reforms, decollectivization of the rural economy has removed the commune shelter and increased household opportunities as well as risks, so that decisions such as marriage are likely even more dependent on calculation of costs and benefits. The prospective husband's location, where the wife will likely spend the rest of her life, is an even more crucial consideration for her than in pre-reform years (Li and Lavely 1995).

Second, spatial restructuring since the reforms has widened the gaps between regions, especially between coastal and inland regions (e.g. Fan 1995; 2005a; Wei 2000). These large regional differences motivate peasant women to seek spatial hypergamy across provinces and long distances (Fan and Huang 1998; Tan 1996b; Wang 1992; Yang 1991). Studies have shown that long-distance marriage migration is a relatively new phenomenon and has increased over time (Tan 1996b; Wang 1992; Xu and Ye 1992; Yang 1991), giving rise to the expression *wailainu* or *wailaimei* (women from outside). A popular saying in rural Zhejiang – a major destination of female marriage migrants – is, "In the 1960s, wives were from Subei [a poor region in the neighboring province of Jiangsu]; in the 1970s, they were from rustication; and in the 1980s, they come from afar" (Xu and Ye 1992). It suggests that marriage migration from poor areas is not a new phenomenon but that long-distance marriage migration is partly explained by economic changes and increased mobility since the 1980s.

The patterns of marriage migration are unduly affected by the hukou system. As discussed in Chapter 5, peasant women's lack of urban hukou and their inferior background render them among the least desirable in the urban marriage market. The options for peasant women desiring spatial hypergamy are, therefore, largely restricted to rural areas.

For long-distance marriage migration, an intermediary that provides information not only about the prospective spouse but also about the specific origin and destination is especially crucial. For example, previous marriage migrants may inform their sisters, relatives, friends, and acquaintances from the native place about the host community and introduce them to prospective husbands (Wang and Hu 1996; Xu and Ye 1992; Yang 1991). The resultant chain or "snowballing" migration would likely result in focused migration streams, from specific origins to specific destinations (Min and Eades 1995). In addition, it is likely that attribute matching and trade-off are emphasized even more among long-distance moves than short-distance moves, because prospective spouses have little else to rely on for information and because the risks are higher than short-distance moves.

Some studies have found that men that marry migrant women are typically older or poorer than other men and some are mentally or physically disabled (Ma *et al.* 1995; Xu and Ye 1992). While their less desirable attributes impose constraints on finding mates locally, they may be considered "matchable" with women from poorer regions. To women desiring spatial hypergamy, a prosperous region is an attribute that can offset the prospective husband's less desirable personal attributes. And their long distance from the natal families may be more than offset by the benefits of a prosperous destination. Migrant women from poor regions are known to expect a lower bride-price and less costly weddings, and are perceived to be diligent workers (Liu 1990; Xu and Ye 1992). To their husbands, these qualities may very well offset the migrant women's less desirable backgrounds.

In addition to relying on acquaintances as intermediaries, prospective spouses may also use the service of brokers and other media. The rise in entrepreneurship, permitted and encouraged by the economic reforms, is conducive to the work of marriage brokers, who collect a fee by arranging for prospective husbands to meet potential marriage partners (Tan 1996b; Yang 1991). Some men who have difficulties finding a wife locally welcome the brokers' service, although incidents of divorce, abduction, and fraud have led to a perception that such a channel is "cheap but risky" (Min and Eades 1995).[62] Other media, such as advertisements and "introduction centers," are also used (Tan 1996b), again reflecting better information flows and rise in entrepreneurship. A popular magazine called *Zhongguo Funu* (*Chinese Women*), for example, regularly advertises for a marriage introduction center. Besides the standard information such as age, height, education level, and economic status, these advertisements often specify urban/rural and location criteria. The following example describes an above-marriageable-age man from a less developed part of Jiangsu (Subei), hoping to marry a peasant woman from another province:

> Miss, do you want to come to the relatively well-off Jiangsu province? Male, aged 30, height 1.65 meters, single, senior high education, loves calligraphy, painting job, healthy, law-abiding, kind and cheerful, annual income 20,000 yuan plus, savings, six-room flat. Looking for a woman from rural, poorer, and mountainous areas as companion. No requirements on location, education level, or marital history. Send letter to Jiangsu province, Xiangshui county.
>
> *Zhongguo Funu* 1997b

It is not uncommon that men's advertisements specify whether they desire urban or rural wives, highlighting location as an important attribute in the matching process. In the above case, the man specifically stresses the relative wealth in Jiangsu, showing that he considers his location an appeal, which may offset his possibly less desirable attributes not specified in the advertisement. Furthermore, targeting women from poorer provinces suggests that he expects that the inferior background of a prospective wife would facilitate a matching relationship. Women's advertisements typically require that the potential spouses be located in more desirable locations than theirs. For example, a woman from Anhui looks for prospective spouse in "more developed regions" (*Zhongguo Funu* 1997c). Another woman from Guizhou, for a long time the poorest province in China, wants to marry someone who can help her migrate and leave that province (*Zhongguo Funu* 1997c).

Labor migration and marriage migration

Labor migration has important implications for marriage and marriage migration in rural China. First, increased labor mobility allows information about regional differentials in development and conditions of prosperous regions to travel more quickly and widely. As rural women now have greater access to information about distant locations, the geographical size of their marriage market also expands.

Consequently, spatial hypergamy may involve not only village-to-village or county-to-county moves but can also be achieved through interprovincial, long-distance migration.

Nevertheless, peasant women have tremendous difficulties entering the urban marriage market. In terms of attribute matching and trade-off, they are among the least competitive in the urban marriage market. As the narratives in Chapter 5 show, peasant women are generally unable to marry urban men unless the latter have an undesirable attribute, such as old age or disability. The story of Wang Xiaoli further illustrates rural women's constraints (*Zhongguo funu* 1997a). She lives in a poor rural area in Sichuan, and was forced to leave school at grade four, so that her younger brother could go to school. For five years, her father has been trying to find a "matchable" husband for Xiaoli, who on the other hand decided to go to Beijing by herself: "I do not want to live like my mother; I must go out and adventure." Upon reaching and working in Beijing, Xiaoli discovered that the only persons willing to marry her were either physically handicapped or 20 years older than she. She concludes, "Only those who cannot find a wife locally [Beijing] would consider outsiders like us. Unless we are very pretty, nobody would pay us any attention. Maybe my father is right: peasants should remain in rural areas" (*Zhongguo funu* 1997a). Xiaoli's experience is consistent with the general sentiment among peasant women that they have few options other than returning to the countryside for marriage.

Second, through migrant work, peasant women have opportunities to meet male migrants from other parts of the country. Thus, labor migration expands the marriage market of migrants to include locations distant from their home villages. The marriages that ensue – *dagong* marriages – are less likely arranged by parents or intermediaries, although they may still involve spatial hypergamy, attribute matching, and trade-off. Information about *dagong* marriages is extremely limited. Narratives in Chapter 6 suggest that peasant women are weary of marrying other migrants due to uncertainty about the latter's backgrounds. Later in this chapter, I shall discuss briefly how *dagong* marriages differ from spatial hypergamous marriages.

Finally, as discussed in Chapter 5, gender division of labor whereby the husband does migrant work and the wife stays in the village is a popular household strategy in rural China. In "Evidence from the Gaozhou field study," I shall illustrate how the split-household strategy boosts the utility of marriage and encourages household formation, even if marriage may entail bringing in a wife from a distant location.

Evidence from the 1990 and 2000 censuses

Marriage accounts for respectively 28.3 percent and 28.9 percent of female intraprovincial and interprovincial migrants, according to the 1990 census (Table 8.1). While the magnitude of interprovincial moves (1.4 million) is, as expected, smaller than that of intraprovincial moves (3.1 million), marriage is

Table 8.1 Comparison between female "industry/business" and "marriage" migrants, 1990 and 2000 censuses

	1990 census						2000 census		
	Intercounty migrants			Interprovincial migrants			Interprovincial migrants		
	Industry/ business	Marriage	All migrants	Industry/ business	Marriage	All migrants	Industry/ business	Marriage	All migrants
Number (million)	2.4	4.4	15.5	0.8	1.4	4.9	9.5	1.6	15.1
% of all migrants	15.4	28.3	–	16.6	28.9	–	62.7	10.5	–
Age									
Mean (years)	25.2	25.6	26.7	25.0	25.4	27.2	25.3	26.9	26.0
15–29 (%)	79.4	87.6	66.3	79.8	86.1	63.0	76.0	79.7	69.1
Marital status: married (15+) (%)	40.1	99.8	63.6	43.3	100.0	67.5	45.2	99.7	54.1
Education (6+) (%)									
Elementary or below	38.4	52.5	41.2	42.2	59.5	46.5	23.7	39.6	29.6
Senior secondary or above	8.0	7.7	25.2	7.2	6.4	20.9	12.4	12.0	18.1
Labor force participation (15+) (%)	98.1	81.3	65.3	97.8	81.8	68.1	95.4	73.7	74.7
Occupation (15+) (%)									
Services	17.7	2.2	9.4	35.5	4.2	29.5	11.2	2.4	10.3
Industry	54.3	7.9	29.4	56.5	6.5	28.4	71.1	9.6	60.4
Agriculture	9.4	85.3	39.7	8.0	89.2	42.1	2.8	79.2	12.6
Origin (%)									
Streets	4.0	3.0	15.2	5.3	3.5	21.4	5.5	9.4	11.7
Towns	10.6	10.8	18.6	8.6	8.3	13.8	52.4	46.1	48.6
(residents' committees)	–	–	–	–	–	–	5.0	7.0	7.0
(villagers' committees)	–	–	–	–	–	–	47.4	39.1	41.5
Townships	85.4	86.2	66.2	86.2	88.2	64.9	42.1	44.5	39.8

continued

Table 8.1 continued

	1990 census						2000 census		
	Intercounty migrants			Interprovincial migrants			Interprovincial migrants		
	Industry/ business	Marriage	All migrants	Industry/ business	Marriage	All migrants	Industry/ business	Marriage	All migrants
Destination (%)									
Cities	61.1	30.0	54.2	47.1	18.0	45.0	48.8	17.9	49.5
Towns	–	–	–	–	–	–	27.8	8.8	22.2
Counties	38.9	70.0	45.8	52.9	82.0	55.0	23.4	73.2	28.3
Hukou type: agricultural (%)	91.4	89.9	57.5	91.9	92.6	60.6	90.8	85.8	80.8
Hukou location: permanent migrants (%)	4.6	56.0	32.1	3.1	53.0	49.5	0.7	55.0	13.5
Interprovincial migration % of intercounty migrants	–	–	–	33.9	32.1	31.5	–	–	–
Average distance (km)	–	–	–	792.3	930.8	901.3	826.4	877.3	865.4

Sources: 1990 census one percent sample; 2000 census 0.1 percent interprovincial migrant sample.

indeed the leading reason for female migration for both interprovincial and intraprovincial moves, suggesting that distance is not a significant deterrent to marriage migration. In the 2000 census, "industry/business" replaces marriage as the leading reason for female interprovincial migration. The sharp increase of "industry/business" migrants between the two censuses (see Chapter 4) explains to a large extent the decline of the proportion of female marriage migrants from 28.9 percent to 10.5 percent. Despite this decline, the number of female interprovincial marriage migrants actually increased from 1.4 million to 1.6 million, which indicates that long-distance marriage migration continues to be prevalent.

Characteristics of female marriage migrants

A comparison of female marriage migrants with female "industry/business" migrants[63] and female migrants as a whole can shed light on the characteristics of female marriage migrants and their migration decisions (Table 8.1). Both female marriage and "industry/business" migrants are young and heavily concentrated in the 15–29 age range. The youth of both types of migrants reflects their life cycles – peasant women marry young and they engage in migrant work primarily before getting married (see also Chapter 4). Indeed, while practically all female marriage migrants are married, the majority of female "industry/business" migrants are single.

Both female marriage and "industry/business" migrants have lower educational attainment than female migrants as a whole. Between marriage and "industry/business" migrants, the former have lower levels of education. According to the 1990 census, respectively 52.5 percent and 59.5 percent of female intercounty and interprovincial marriage migrants have elementary or below levels of education, compared to 38.4 percent and 42.2 percent for "industry/business" migrants. By the 2000 census, improvement of education reduces the proportions for all groups, but female interprovincial marriage migrants continue to have a higher percentage – 39.6 percent – than female interprovincial "industry/business" migrants and all female interprovincial migrants.

Despite female marriage migrants' low educational attainment, their labor-force participation rates (at the destination) are high – 81.3 percent for intercounty migrants and 81.8 percent for interprovincial migrants based on the 1990 census, and 73.7 percent for interprovincial migrants based on the 2000 census. Although these rates are not as high as those of female "industry/business" migrants, they do indicate that female marriage migrants are active participants in the host economy, contradicting the conventional view that defines female marriage migrants as tied movers rather than economic agents. The occupational structure of female marriage migrants also stands out. The vast majority of them engage in agriculture – respectively 85.3 percent and 89.2 percent for intercounty and interprovincial migrants according to the 1990 census and 79.2 percent for interprovincial migrants in the 2000 census – whereas the majority of "industry/business" migrants work in the industrial sector.

The above characteristics are consistent with migrants' origins and destinations. The proportions of female marriage and "industry/business" migrants originating from rural areas – townships in the 1990 census and townships and villagers' committees combined in the 2000 census – are higher than those of female migrants as a whole. Both groups also have very high percentages with agricultural hukou classification. The predominantly rural origins of both female marriage and "industry/business" migrants explain partly their low educational attainment. The two groups, however, differ in their destinations. The vast majority of marriage migrants – in the range of 70 percent to more than 80 percent – move to counties. Considerably smaller proportions of "industry/business" migrants move to counties, and the majority of them – except in the case of interprovincial migrants in the 1990 census – move to cities or towns. Thus, not only are female marriage migrants largely from rural areas, the bulk of them choose also to migrate to rural areas, which in part explains their high concentration in agriculture. The prominence of rural destinations supports the argument that peasant women's marriage market is predominantly confined to rural areas.

Another difference between female marriage and "industry/business" migrants is that the majority of the former but only minute proportions of the latter are permanent migrants. As discussed in Chapter 4, "industry/business" migrants are mostly denied urban hukou, whereas marriage migrants to rural areas are generally able to obtain hukou at the destination. Yet, the proportions of female marriage migrants that change their hukou to the destination – in the order of about 55 percent – appear quite low. This may signify that some female marriage migrants to rural destinations still do not enjoy local resources as much as local residents. In theory, allocation of contract land is based on household size. In practice, and because land reallocation happens only infrequently, the arrival of a bride does not necessarily increase the amount of farmland for the household. In the 2005 Sichuan and Anhui Interview Records, for example, most villagers indicate that land reallocations have not occurred for at least ten years. A 32-year-old respondent, for example, indicates that his and his brother's land allocation remains the same despite the arrival of their wives and the birth of children during the past ten years: "During the past ten years, the village has not had any reallocation of land. Even though our family has grown in size, our farmland has not increased" (2005 Sichuan and Anhui Interview Records).

The above comparisons show that female marriage migrants are a distinct group of women. Most are young and poorly educated, come from rural areas and move to rural areas, are economically active in the labor force, and are mostly engaged in agriculture. They do not fit the profile of family-oriented, passive, and tied movers typically associated with marriage migrants and women moving within marriage. Rather, these peasant women seek to migrate to the destination *and* join the agricultural labor force there. The characteristics of female intercounty and interprovincial marriage migrants, as reported by the 1990 census, are strikingly similar. These characteristics are largely repeated among female interprovincial marriage migrants, as are their differences from female interprovincial "industry/business" migrants, in the 2000 census. What the above suggests is that peasant women with the most severe human capital and locational constraints are most prone to resort to marriage migration. They have very limited opportunities for social and economic mobility other than marriage, and their willingness to move long distances and be away from familiar environments can be interpreted as a price they are willing to pay in order to overcome these constraints.

By moving from one rural area to another and from agricultural work to agricultural work, it seems that female marriage migrants have not accomplished significant advancement in social or economic status. However, their choice of destination and occupation reflects a combination of personal, social, institutional, and locational factors and may indeed constitute an improvement in wellbeing. First, these migrants' low skills, as reflected by their low educational attainment, adversely affect their competitiveness in urban industrial sectors and in cities. Second, as discussed earlier, both institutional and social constraints block peasant women from the urban marriage market. Thus, moving to urban areas may not be an option for them at all. Third, all rural areas are not the same. Those near big cities or in more developed provinces are more desirable, especially since economic reforms have accelerated commercialization of agriculture, enabling farmers near urban markets and in high-productivity regions to seek greater profit. Therefore, marrying someone in another rural area, one that affords better economic opportunities, is an appealing strategy to enhance a peasant woman's economic status and is one that is much more feasible than marrying an urbanite. Finally, regardless of whether they are marriage migrants, rural married women are expected to stay in the village to farm, while their husbands may pursue migrant work. This gender division of labor, discussed in Chapters 5 and 6, means that female marriage migrants who move to rural areas, similar to local married women, are overwhelmingly engaged in agriculture.

The husbands of female marriage migrants

Information about the husbands of female marriage migrants can illustrate further the role of attribute matching and trade-off in marriage migration. While the 2000 census migrant sample does not include migrants' spouses, the 1990 census one percent sample has some relevant information. The latter allows matching of household members into husbands and wives when either is the household head. Since men are much more likely than women to be household heads, I have selected to examine male household heads who are married to interprovincial marriage migrants, interprovincial non-marriage migrants, and non-migrants. In order to control for the effects of age on education, Table 8.2 includes only husbands aged 25–39. This age range accounts for 45 percent of all the husbands of female interprovincial marriage migrants.

Husbands of marriage migrants have the lowest educational attainment, with 42.5 percent with elementary or below level, compared to respectively

150 *Marriage and marriage migration*

Table 8.2 Husbands of interprovincial marriage and non-marriage migrants, 1990 census

	Marriage migrants	Non-marriage migrants	Non-migrants
Education (%)			
Elementary or below	42.5	23.4	34.0
Occupation (%)			
Services	6.7	39.5	13.8
Industry	17.5	43.3	18.6
Agriculture	75.7	17.2	67.6

Source: 1990 census one percent sample.

Note
Only husbands who were household heads and between 25 and 39 years old in 1990 are included.

23.4 percent and 34.0 percent of husbands of non-marriage migrants and non-migrants. Among the three groups, husbands of marriage migrants are most concentrated in agriculture, while most husbands of non-marriage migrants work in industrial or service sectors. As a whole, therefore, husbands of marriage migrants are of lower socioeconomic status than not only husbands of non-marriage migrants but also husbands of non-migrants. These differences support the argument that men that are not competitive in the local marriage market are more likely than other men to seek brides from elsewhere, including those from other provinces.[64]

Spatial patterns of female interprovincial marriage migration

Female interprovincial marriage migration exhibits distinct spatial patterns. Based on the 1990 and 2000 censuses, respectively 84.8 percent and 77.2 percent of all female interprovincial marriage migrants originate from the central and western regions, and respectively 60 percent and 61.6 percent move to the eastern region. Figures 8.1 and 8.2, which illustrate the 15 largest net flows between pairs of provinces, reinforce the observation that female marriage migration is primarily an eastward phenomenon.

The 1990 census figures show that, except for the flows from Hebei to Beijing and Tianjin, the largest net flows of marriage migration originate exclusively from the southwestern provinces of Yunnan, Guizhou, Guangxi and Sichuan, while the net migration to the six eastern coastal provinces of Hebei, Shandong, Jiangsu, Zhejiang, Fujian, and Guangdong accounts for more than 50 percent of the total net flows. Most of the receiving provinces are among the most developed provinces in China, which is clearly a major factor of marriage-migration destination. The net flow from Guangxi to Guangdong is the largest in the nation. Except for the flows from Guangxi to Guangdong, and from Hebei to Beijing and Tianjin, all other flows are between non-adjacent provinces. In fact, the majority of interprovincial flows are across several provinces and involve long distances.

By the 2000 census, the overall patterns of eastward flows, net migration from

Marriage and marriage migration   151

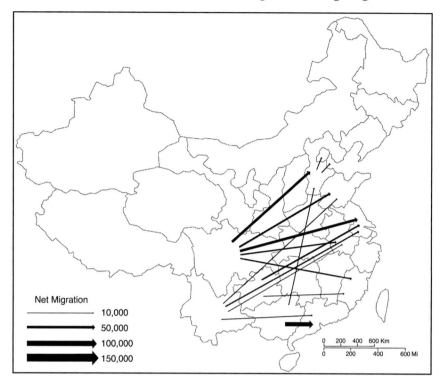

Figure 8.1 Female net interprovincial marriage migration, 1990 census (source: 1990 census one percent sample).

less developed provinces to more developed provinces, and long-distance migration, persist (see also Ding *et al.* 2005). Compared to ten years ago, there are two notable changes. First, while Yunnan, Guizhou, Guangxi, and Sichuan still constitute the majority of the largest donor provinces, the origins of marriage migrants are more diverse than those in the previous decade. New members of the largest donor provinces include three provinces in Central China region – Anhui, Jiangxi, and Hunan – and two in northern and northeastern China – Inner Mongolia and Heilongjiang. Increased prominence of the central region as a donor of marriage migrants reinforces the observations made in Chapter 2 on accelerated out-migration from that region and the economic difficulties faced by Anhui, Jiangxi, and Hunan. Second, Sichuan is the destination of two of the largest net female marriage migration flows, respectively from Yunnan and Guizhou. All other destinations are in the eastern coastal region and are more developed provinces. Sichuan's inclusion into the most prominent destinations likely reflects its widening gaps in economic development from Yunnan and Guizhou (see Table 2.4).

Elsewhere, I have shown that within the receiving provinces, the effects of marriage migration are especially felt in poorer, rural and more peripheral locations (Fan 1999). In Guangdong, female marriage in-migration is more

152 *Marriage and marriage migration*

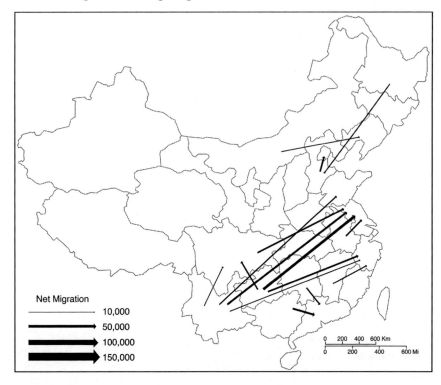

Figure 8.2 Female net interprovincial marriage migration, 2000 census (source: 2000 census 0.1 percent interprovincial migrant sample).

prominent in the peripheries than in the economic core and is more notable in rural areas than in cities. Western Guangdong – the site for the Gaozhou field study – in particular, distinguishes itself by having more than 80 percent of its female in-migrants arriving as marriage migrants. Another important recipient province, Jiangsu, exhibits similar spatial patterns. Fan and Huang (1998) document that female marriage migrants are more highly represented in Jiangsu's poorer periphery and rural locations than in its economic core and cities.

The spatial patterns of female marriage migration described above support two arguments. First, female marriage migrants seek to move up the spatial hierarchy, from inland to coastal provinces and from poorer to more developed regions. In the Chinese-language literature, this primarily eastward movement is well summarized by the phrase "spring water flowing east" (Zhang and Zhang 1996). These patterns reinforce the notion that marriage migration is an economic strategy. Second, the concentration of origins and destinations depicts distinct migration streams, supporting the argument that social and kinship networks play an important role in facilitating and sustaining marriage migration, including long-distance marriage migration. These arguments will be substantiated further by the Gaozhou field study.

Evidence from the Gaozhou field study

For the purpose of examining further household and individual-level processes, considerations and decisions in relation to marriage migration, the rest of the chapter draws from the 1999 Gaozhou field study (see Chapter 1). Gaozhou is a county in western Guangdong (Figure 8.3).

The field site

Western Guangdong is a popular destination of marriage migrants and a prominent donor of labor migrants. According to the 1990 census, marriage accounts for more than 80 percent of all female intercounty migrants to six western Guangdong counties (GDPPCO 1992) – Wuchuan, Gaozhou, Xinyi, Dianbai, Huazhou, and Luoding – all heavily agricultural and some distance away from the province's most thriving economic centers such as Dongguan and Shenzhen in the Pearl River Delta (Figure 8.3). Among the six counties, Gaozhou[65] is ranked first in the number of female marriage in-migrants. Gaozhou is also characterized by a high proportion of labor out-migrants mostly working in the Pearl River Delta.

Within Gaozhou, Genzhi township is selected as the field site because its demographic and economic conditions are near the average for Gaozhou. In 1996, Genzhi's population was 61,703, compared with an average township population of 51,849 in Gaozhou. In the same year, Genzhi's per capita rural output was 4,910 yuan, compared with 4,368 yuan for Gaozhou county as a whole. Like many other townships in Gaozhou, Genzhi is heavily dependent on lychee trees, which are grown on three-quarters of the township's hilly land. According to the township authority, female marriage migrants are represented

Figure 8.3 Site of Gaozhou field study and origins of marriage migrants (source: 1999 Gaozhou field study; reproduced from Fan and Li (2002) with permission).

154 *Marriage and marriage migration*

in approximately one-fifth of the households, and more than one-third of the township's labor force is engaged in migrant work.

Within Genzhi, and with the help of local informants, I selected two adjacent villages, which for confidentiality purposes are referred to as "Village A" and "Village B." In terms of overall economic conditions, the two villages are near the average for Genzhi. In 1997, the per capita income of these two villages was 5,003 yuan, whereas the per capita income in Genzhi was 5,018 yuan. According to Genzhi's registration records, as of 1998, Villages A and B consisted of a total of respectively 128 households and 602 residents. Like most traditional villages, they exhibit strong family lineage. Two family names account for respectively 59 percent and 24 percent of the households in Village A, and one family name represents 83 percent of the households in Village B.

I conducted a questionnaire survey of the villages and taped interviews with village residents, village and township authorities, and local informants in January, February, and December of 1999. Most of the informants are cadres at the Gaozhou Statistical Bureau, Gaozhou Family Planning Commission, and Genzhi Family Planning Commission. The survey had a response rate of 60 percent and included a total of 76 households. I also selected residents from a range of backgrounds for in-depth interviews on topics related to marriage, marriage migration, and migrant work.

Origins and intermediaries of marriage migrants

The wives in about one-third (25) of the households surveyed are marriage migrants. Table 8.3 compares migrant wives from different origins. Six are from other counties within Guangdong, 14 from Guangxi province, and five from other provinces – two from Hunan and one each from Yunnan, Henan, and Jiangsu. The prominence of non-Guangdong origins, which account for 76 percent of the migrant wives surveyed, underscores the importance of long-distance marriage migration. Relatives and *tongxiang* are the most prominent means for Gaozhou husbands to meet wives from other provinces, accounting for 86 percent of the marriages with migrant women from Guangxi and 75 percent of the marriages with migrant women from the other four provinces. For marriages with Guangdong wives, 40 percent are attributable to relatives and *tongxiang* and another 40 percent involve husbands and wives meeting each other during migrant work.

All the Guangdong wives are from counties near Gaozhou – Dianbai, Yangchun, and Luoding – suggesting that their moves are variants of traditional short-distance marriage migration (Figure 8.3). In contrast, all the wives from Guangxi are from origins more than 300 km away and relatively near to the provincial capital of Nanning city. Specifically, six Guangxi wives are from Yongning county, three from Chongzuo county, one from Daxin county, one from Mashan county, one from Dahua county, and two from unspecified locations near Nanning city. These moves are, by the standard in rural China, long-distance moves (see also Chapter 4). The origins' relative proximity to the provincial capital has probably facilitated contacts with visitors and access to

Table 8.3 Marriage intermediaries, 1999 Gaozhou field study

	All migrant wives	Origin		
		Guangdong	Guangxi	Other provinces
Total number of respondents	25	6	14	5
How did husbands and wives meet? (%)				
Relatives and *tongxiang*	74	40	86	75
During migrant work *(dagong)*	22	40	14	25
Family members	4	20	0	0
Other	0	0	0	0
N	23	5	14	4
Marriage migrant acquaintances in Gaozhou (average number)				
Tongxiang	1.2	0.0	2.1	0.2
Relatives	0.2	0.2	0.1	0.4
Family members	0.1	0.0	0.2	0.0
N	25	6	14	5

Sources: 1999 Gaozhou field study; adapted from Fan and Li (2002) with permission.

Note
N refers to the total number of responses.

information about other provinces. In addition, Cantonese (*baihua*), the most common dialect in Guangdong, is popular in Guangxi, which enables frequent interactions between the two provinces. A migrant wife from Chongzuo recalls: "Many people from Guangdong did business in Chongzuo, including selling longan. That's how my cousin was introduced to her husband and moved to Gaozhou. Later, my cousin brought me here and introduced me to my husband."

The above account is similar to the stories of several other marriage migrants, who first learned of Gaozhou because of a relative or *tongxiang* who had migrated there, and after visiting Gaozhou themselves became attracted to the idea of meeting and marrying men there. All the three migrant wives from Chongzuo share the same family name and came to Gaozhou between 1989 and 1993. Similarly, four of the six migrants from Yongning who share a family name all moved to Gaozhou between 1986 and 1989. The sharing of family name and the timing of their moves suggest that they are relatives or *tongxiang* that are part of chain marriage migration. The 14 migrants from Guangxi have a total of 29 *tongxiang*, three sisters and two other relatives who also married Gaozhou men (Table 8.3). On average, every Guangxi wife knows of 2.1 *tongxiang* who are also migrant wives in Gaozhou. This contrasts with averages of 0 *tongxiang* for Guangdong wives and 0.2 *tongxiang* for wives from other provinces. The larger number of acquaintances among Guangxi wives is strong evidence of chain migration and underscores the role of social and kin networks in long-distance marriage migration.

Matching, trade-off, and location

A common perception in the two villages is that the large economic gap between Gaozhou and the migrant wives' home villages is key to their choosing to come. A migrant wife from Guangxi comments: "My *tongxiang* told me that this was a good place, that this was a much better place than my home village. They said that life here would not be as bitter" (Zhao Xiujuan[66] – a marriage migrant from Guangxi, author's interview in December 1999).

In particular, Gaozhou's resources of fruit trees have been quite profitable and are a notable attraction to marriage migrants, as this woman from Guangxi describes: "Life is better here because we can make money from fruit trees. In Guangxi we have farmland but no fruit trees" (author's interview in December 1999). According to my informants, the increased productivity of fruit trees since the mid-1980s has improved significantly the well-being of village residents and made possible the building of new houses, sometimes referred to as "lychee houses." Indeed, both villages exhibit a mosaic of big and newly built residences along with older, run-down houses (see Plate 8.1).

Field observations further suggest that good location can offset less-desirable attributes. Table 8.4 summarizes the age and education levels of migrant wives, non-migrant wives, and their husbands. In order to focus on similar cohorts and minimize the effect of generational differences, the comparison is limited to those who married since 1983, which includes all 25 migrant wives and 17 (35 percent of surveyed) non-migrant wives.

Plate 8.1 Villages in Gaozhou are characterized by a mix of older houses and new and bigger "lychee houses." (source: Author, Gaozhou, January 1999).

Table 8.4 Age and education of wives and husbands, 1999 Gaozhou field study

	Migrant wife				Non-migrant wife*
	All	Origin			
		Guangdong	Guangxi	Other	
Number of respondents	25	6	14	5	17
Mean age at marriage					
Wife	21.6	21.8	21.6	21.5	22.1
Husband	26.8	26.4	26.1	29.8	25.2
Husband–wife age gap (mean years)	4.8	3.8	4.5	6.8	3.1
Education (% elementary or below)					
Wife	44	33	43	60	53
Husband	60	50	50	100	71
Husband–wife education gap (%)					
Same	54	67	46	60	71
Husband has higher education	33	33	31	40	24
Wife has higher education	13	0	23	0	6

Sources: 1999 Gaozhou field study; adapted from Fan and Li (2002) with permission.

Note
* Only those married since 1983 are included.

Not surprisingly, across all groups, husbands are on average a few years older than their wives. The age gap ranges from −2 (wife older by two years) to 19, and the mean age gap is larger between migrant wives and their husbands (4.8) than between non-migrant wives and their husbands (3.1). The husband–wife age gap varies with migrant origin – wives from Guangdong have the lowest mean age gap (3.8), followed by those from Guangxi (4.5) and other provinces (6.8), from their husbands. All four couples that have age differences of ten years or more involve marriage migration – two wives are from Guangxi, one from an adjacent county (Dianbai), and the fourth is from Hunan. Among these four couples, the wives' ages ranged from 17 to 23, and the husbands' ages ranged from 29 to 34, at the time of marriage. These data are consistent with many respondents' perception that older men who have difficulties finding mates locally will resort to marrying migrant women, and that women from poorer areas are more willing than local women to accept larger age differences, in light of the wives' inferior background and the husbands' favorable location.[67]

The majority of women surveyed married men with similar or higher levels of education, and the typical combination consists of a husband with junior secondary and a wife with elementary education. However, three Guangxi wives (23 percent) have higher educational attainment than their husbands, which is not common in Chinese society. All three husbands have elementary level education while the wives' education levels range from junior secondary to senior secondary. This anomaly suggests that the husbands' lower educational attainment is offset by their favorable location, especially to migrant wives from poor locations. This trade-off is similar to the age-gap trade-off described earlier. In fact, all three Guangxi wives that have higher educational attainment than their husbands are also considerably younger than their husbands, with age gaps of respectively four, six, and ten years, further supporting the argument that location is an important consideration in the trade-off exercise. In addition, husbands of migrant wives have lower educational attainment than husbands of non-migrant wives – 40 percent of the former and 29 percent of the former have elementary or below level of education. Similar to the effect of older age, it appears that men with lower educational attainment are less competitive in the local marriage market and are more likely to marry migrant wives.

When asked, "Why do you think local men choose to marry women from other places?" 45 percent of village respondents refer to men's attributes, 27 percent to migrant work that increases opportunities to meet non-local women, 12 percent to the perception that migrant wives are more hardworking, 10 percent to the high sex ratio in Gaozhou due to out-migration of young women, and 7 percent to less expensive weddings. When asked to elaborate on their answers, most respondents mention age (difference) and economic factors. Migrant wives are expected not only to contribute to economic production but also to be hardworking. And, migrant wives are known to not anticipate costly weddings or lucrative bride-price. Both of the above are practical and economic considerations on the part of the husbands' families. Indeed, the average wedding cost (including bride-price and adjusted for inflation) is significantly different between marriages with local women (7,141 yuan) and those with migrant wives (4,619 yuan). Weddings with wives from Guangxi were the least expensive, averaging only 3,932 yuan.

The above observations underscore the importance of economic factors in the matching and trade-off process. Men with low economic ability have difficulties finding mates locally and are motivated to marry women from poorer provinces, and at the same time these women are willing to accept their husbands' low economic status (relative to other men in the destination) because of their favorable location. A man 45 years old who is poor by the villages' standard and who married a migrant woman more than ten years younger told us:

> I was introduced to my wife through friends .... We didn't have a wedding banquet. No local women would marry me. I had been introduced to many women ... but none worked out because they asked for too much money.

The account of Zhao Xiujuan – a 32-year-old marriage migrant from Guangxi – further illustrates the workings of attribute matching and trade-off. Xiujuan was born in a village near Nanning. She learned about Village A and her husband through her cousin, who married a man in Village A three years before Xiujuan came. She was married at the age of 21. When asked about marriage decision-making, Xiujuan comments:

> In this village, my husband was considered rather poor. No one wanted to marry him [laughing]. Before me, he was introduced to other women, including one from Hainan [also a poorer province], but they came and noticed that he didn't have much [in the house] .... I thought we matched quite well – he had junior secondary education and I had not even finished elementary school. He is five years older than I.

In the survey, the majority (55 percent) of village respondents indicate that "solving men's mate-finding problems" is the most significant contribution of marriage migrants. As aptly summarized by one local informant:

> Men usually don't care much about the family background of their wives.... They are more concerned with their wives' appearance. That's why there is an old saying that "it's easier to pick a wife than to pick a husband." Women look at not only men's appearance but also their family and where they live.

All in all, the field observations summarized above highlight the importance of location as a marriage attribute, and are evidence for the trade-off between location on one hand and age, education, and economic ability on the other. This trade-off, in turn, motivates spatial hypergamous migration.

Marriage and labor migration

The field site, which is characterized by a high degree of labor out-migration, is uniquely suitable for studying the relationship between migrant work on one hand and marriage and marriage migration on the other.

Dagong marriage migration

Labor migration or *dagong* (see Chapter 1) increases the opportunities for single men and women to meet prospective mates from other places. In the survey, 22 percent of marriage migrants met their husbands during migrant work (see Table 8.3). Unlike spatial hypergamous marriages described earlier, these "*dagong* marriages" appear less dependent on attribute trade-off, as described by this informant: "They [*dagong* marriage migrants] have a different kind of marriage ... they met their husbands during migrant work, and their marriages are built on affection".

The account of Lin Xiaoying supports the notion that location is less important in *dagong* marriage migration. Xiaoying was born in Chenzhou, Hunan, near the northern border of Guangdong. She has the same level of education as her husband – junior secondary – and is one year younger than her husband. They met while working in Shenzhen, and she was married at

160 *Marriage and marriage migration*

the age of 22. When comparing Village A with her home village, Xiaoying comments:

> They [her acquaintances] all said that Guangdong was better. But back in Hunan there was less farm work – we grew two rice crops a year and could rest in the winter. Here, farm work is busy; in the winter we grow vegetables as well.

She expects her younger sister to be close to the natal family: "My younger sister should marry someone near home so she can take care of our parents".

The next section shows that spatial hypergamous and *dagong* marriage migrations differ also in terms of gender division of labor.

Gender division of labor

More than half (52 percent) of the husbands of marriage migrants (13 out of 25) and 41 percent (7 out of 17) of the husbands of non-migrant wives are engaging in migrant work, almost exclusively in the Pearl River Delta. All send home remittances, ranging from 900 yuan to 40,000 yuan annually, with an average of 6,713 yuan. Given a mean household income of 9,861 yuan in these two villages, these remittances are indeed a major source of household income.

The split-household strategy is amply illustrated in the two villages. Many respondents comment that labor migrants would eventually return and use their savings to help build new houses. While the husband works as a labor migrant, the wife assumes all village responsibilities. In the survey, 19 of the 20 households where husbands are out-migrants have children younger than 10 years old. The vast majority (80 percent, or 16 out of 20) are two-generation households, where grandparents are not available to help.[68] The wives have therefore become the primary persons responsible for not only agricultural work but also house chores and raising young children (see also Plate 8.1).

Among the five *dagong* marriages in the field study, however, only one of the husbands continues migrant work after he was married. Though the sample is too small for assessing if the split-household strategy is less prevalent among *dagong* marriage households, the role of affection may be a factor in their decision-making, as described by a local informant: "Their [women who met their husbands during migrant work] marriages are built on affection, and so the husbands and wives want to stay together". For example, Lin Xiaoying, whom I introduced earlier, and her husband returned to Gaozhou after they got married. Both are now engaging in farm work, growing fruit trees and raising domestic animals.

Summary and conclusion

The migration literature has largely ignored marriage migration and the research on marriage migration has focused primarily on international marriages and capitalist economies. In addition to documenting marriage migration within China, this chapter has challenged the conventional assumption of a unidirectional relationship between marriage and migration and the notion that marriage migrants are passive actors whose mobility is simply a response to marriage. Instead, I have argued that marriage is an important means for peasant women to achieve mobility and exert their agency in advancing their well-being.

The large volume of interprovincial eastward marriage migrants in recent decades indicates that economic restructuring has enlarged the geographical size of the marriage market and motivated peasant women's pursuit of spatial hypergamy. Analyses of census data show that rural women in inferior institutional, economic, and human capital positions are most likely to resort to marriage migration. To them, marriage is much more than a life event but is a valuable means to move to a more prosperous location. Without the skills necessary to compete in the urban labor market, these women are not part of the rural–urban labor migrant regime. And, they are excluded from the urban marriage market. Most marriage migrants are, instead, attracted to rural areas in well-developed regions and provinces.

The tradition of attribute matching and trade-off continues to dominate marriage decision-making in China. Traditional marriages that involve the migration of the bride to the groom's family in a nearby village remain the dominant mode in the countryside. However, as the regional gaps of development widen, and as increased mobility facilitates the travel of information about regional inequality, location has emerged as an important factor in marriage decision-making. Through the help of intermediaries such as fellow villagers and relatives, poor peasant women can pursue spatial hypergamous marriages that may involve long-distance migration to more prosperous rural areas. The prevalence of chain marriage migration is a major explanation for distinct marriage migration streams from specific origins to specific destinations. At the same time, men that are disadvantaged in the local marriage market are being matched with migrant wives from poorer regions, illustrating the trade-off between location on one hand and age, education, and economic ability on the other.

I have also shown that labor migration has important implications for marriage. First, labor migration has widened peasants' social spaces and expanded the geographical size of their marriage market. There is some evidence, albeit anecdotal, that *dagong* marriages are more solidly grounded than spatial hypergamous marriages in affection. Second, labor migration necessitates a split-household strategy and gender division of labor, a subject also discussed in earlier chapters.

The important role of location in marriage decision-making, the matching between husbands and wives, the interactions between marriage and labor migration, and the gender division of labor within marriage all highlight household and individual-level strategies in response to macro-level constraints and opportunities. These strategies underscore the importance of agency and the centrality of marriage for understanding migration.

9 The Chinese migrant in the twenty-first century

Introduction

During the first decade of the new millennium, "China's rise" and "China's century" are among the descriptions favored by media and observers worldwide to portray the nation's rapid transformation into a global economic superpower (e.g. Bader 2006; *Newsweek* 2005). China is now the third-largest trading nation in the world. According to the WTO, China has since 2006 surpassed the US as the world's second-largest exporter and is projected to overtake Germany to become the biggest exporter in 2008 (*China Daily* 2007c). And, all eyes will be on China in August 2008 as the Olympic Games opens in Beijing, and in May–October 2010 as Shanghai hosts the World Expo. Behind the glamour, millions of migrant workers have been toiling for decades to make, build, and serve. It is estimated that Beijing alone has about four million peasant migrants, of whom a quarter are construction workers, many brought in to build the Olympics facilities (Xinhua News Agency 2006).[69]

In this concluding chapter, I raise three questions that are central to understanding, interpreting, and perhaps projecting migration, especially rural–urban migration, in China at the beginning of the twenty-first century. First, will the mobility increase observed for the last decade or so continue in the foreseeable future? Second, with more than two decades of migrant work experiences and as these experiences are passed to the next generation, are Chinese peasants developing effective strategies to enhance their positions? Answers to both questions will also inform analysis of migrant labor shortage. Third, how are state policies, including hukou reforms, approaching the question of the Chinese migrant? All three are "big" questions each warranting exhaustive treatment. This chapter aims at providing some brief thoughts and observations about these questions, drawing upon recent news reports and preliminary findings from the 2005 Sichuan and Anhui Interview Records (see Chapter 1). I close the chapter by summary remarks on the hukou and permanent migrant paradigms.

Mobility

The Chinese population is, no doubt, increasingly mobile. Migrants moving for economic betterment and driven by market forces are increasingly defining population movements in China (Poncet 2006; see Chapter 4). The floating population doubled in size between the mid-1990s and 2005 and according to one estimate is increasing by about five million every year (Beijing sheke guihua 2000). Based on this estimate and the official number of 150 million in 2005, by the years 2015 and 2025 the floating population will amount to, respectively, 200 million and 250 million and will account for 14 percent and 17 percent of China's population (NBS 2006; United Nations 2007). Eighty percent of the floating population are peasant migrant workers (Da jiyuan 2006). Migration has, clearly, accelerated the level of urbanization and the effect is especially felt in large cities. One-third of the 17 million people in Shanghai are migrants (Da jiyuan 2006).

This trend of mobility increase refutes the notion that migrant labor is drying up, despite recent reports of labor shortage in southern China (see below). On the contrary, labor migration has firmly established itself as a way of life, perhaps even a culture, throughout China's countryside (Lee 2007: 204; Zhao 1998). Out-migration has become a necessary, desirable source of livelihood and employment for many rural Chinese and a key instrument to relieve agricultural labor surplus. Agricultural labor surplus is large to begin with and is expected to increase for the following reasons (see Chapter 1). First, most analysts agree that China's joining the WTO has adversely affected agriculture (CASS 2003), which exacerbates labor surplus in rural areas. Second, the desire to leave agriculture is strong, especially among young rural Chinese (Croll and Huang 1997; Wang 2003). And third, non-agricultural employment opportunities in rural areas, such as those offered by township–village enterprises (TVE), are saturated and not expected to grow (Johnson 2002).

Villagers in the 2005 Sichuan and Anhui Interview Records overwhelmingly refer to migrant work as an important means to improve their economic well-being. This 36-year-old Sichuan woman, who stayed in the village throughout the ten years or so during which her husband did migrant work in Shenzhen and other cities, summarizes succinctly: "Our primary source of income is migrant work. Agriculture is secondary" (2005 Sichuan and Anhui Interview Records).

The vast majority of migrants in the 2005 Sichuan and Anhui Interview Records continued migrant work in one form or another throughout the ten years since last interviewed. While some had returned to farm for a period of time before going out again (see "Circulation") and others returned because of family needs or difficulties finding urban jobs, almost no respondents returned because of economic opportunities in the countryside.

Migrant work as a means of livelihood is also being passed down from one generation to the next. Parents who engaged in migrant work in the 1980s and 1990s may have returned, while their children are now old enough to go out and send remittances back to support the family. A 47-year-old villager describes the migrant work history of his family:

> I was injured during migrant work [in a coal mine] in 1990 and became paralyzed. My wife's farming then became the only source of livelihood, until my two children were old enough to go out. Since my children began

migrant work, we have had enough food to eat and our lives have improved a lot because of the money they sent back. My oldest son went out when he was 14 years old. My second son went out when he was 17.

2005 Sichuan and Anhui Interview Records

Peasants' continued and cross-generational desire to engage in migrant work, as illustrated by the above example, also fosters the replenishment of migrant workers in urban areas. This, as discussed in Chapter 6, suppresses wage hikes and partly explains the persistent exploitation in the migrant labor regime (Lin 1997: 175; Qiu et al. 2004).

Some researchers observe that a new generation of peasant migrants is emerging. These young migrants prefer non-agricultural employment to farming, have had little farming experience, are more educated than the older generation of migrants, and are more determined to put down roots in the city (Lee 2007: 206; Qiu et al. 2004; Wang 2003). Like their predecessors, they engage in migrant work in order to improve their economic well-being. It is uncertain, however, if they will eventually return or if they will be able to obtain urban hukou and stay in urban areas (see "Circulation" and "Policy"). Preliminary findings from the 2005 Sichuan and Anhui Interview Records show that many young peasant migrants, more than migrants ten years ago, choose not to return for the Spring Festival. This may be because of changing views about traditional celebrations, difficulties and cost of transportation, young migrants' increased sense of belonging in the city, or some other reasons. And, some studies find that family migration – peasant migrants bringing their spouses and children along rather than leaving them in the countryside – is on the rise (Zhong 2000: 221; Zhou 2004).[70] Nonetheless, a massive, permanent departure from the countryside has not occurred, whereas circular migration continues to be the dominant mode among peasant migrants.

Strategies

Circulation

Circular migration is no longer just a temporary solution but has become a long-term practice of many rural Chinese. Two forms of circulation are taking place, often simultaneously. The first form is the going back and forth between the origin and the place of migrant work, and the second is the circulation among different places of migrant work.

Circulation between origin and destination

Peasant migrants return during the Spring Festival to maintain valued traditions in the family and community; they also return during planting and harvesting seasons in order to support agriculture. The prevalence of this type of seasonal migration has led to the term "wild geese households" (*yan hu*), which compares peasant migrants to wild geese that fly to the south in the fall and the north in springtime. In addition to seasonal migration, many peasant migrants have traveled back and forth between the village and places of migrant work, sometimes staying home to farm for an extended period before going out again. A veteran migrant construction worker gives an account of his migration history:

> I began migrant work in 1983. After getting married in 1984, I stayed home to farm for two years. In 1987, I worked in a coal mine. Then, between 1988 and 1990, I stayed home to farm and build a house. I began migrant work again in 1991. Every year, I returned home before the Spring Festival and helped with planting before going out again. In 2004, I stayed home to farm. My wife went out that year.
>
> 2005 Sichuan and Anhui Interview Records

This migrant worker (and his wife) has made a number of mobility changes over an 11-year period: going out, returning to farm, changing migrant work, returning to farm again, going out, returning to farm, and the wife going out. This example underscores the role of household strategy, which requires that migrants and their family members be flexible in terms of what they do and where they work, that they engage in economic production both in the countryside and via migrant work, and that they maintain the economic and social infrastructure of a rural livelihood.

By tapping into resources in both the city and the countryside, peasant migrants may, and some do, obtain the best of both worlds. This long-time migrant worker gives a succinct and thoughtful analysis:

> Comparing urban and rural people – assuming a four-person household in the city, and if there are one or more laid-off workers in the family, then the financial burden is very heavy. For a four-person household in the countryside, assuming one person is doing migrant work, even if the work is not profitable, there is still the land and the house to support the wife and children. Take our family as an example; our expenses are much lower than urban families. Similar to workers in the city, I make several hundred yuan a month, but I have much more freedom and probably better quality of life than they.
>
> 2005 Sichuan and Anhui Interview Records

He pinpoints three reasons why the countryside is preferred to the city as a permanent place to stay. First, it is much cheaper to live in the countryside than in the city. By engaging in migrant work and sending remittances home, the migrant benefits not only from higher wages but also the rural–urban cost-of-living differentials. Second, the farmland and the house not only serve the subsistence purpose but also constitute a form of insurance. Should migrant work fail, one can still return to farming and has a place to stay in the village. Third, the city is viewed by many peasant migrants as a demanding and difficult place to live.

While this perception is, no doubt, a function of their experiences as migrant workers (Chapter 6), the countryside does have more space, better air quality, and a more relaxing lifestyle.

In addition to financial considerations, the identity that most rural Chinese inherit is one that defines them as permanent peasants. Zhong (2000) refers to rural labor migrants as a "hybrid" class that straddles peasantry and the city (see Chapter 4). While peasant migrants may be playing double roles as both urban workers and farmers, most still consider themselves peasants rather than having double (peasant–urbanite) identities (Lee 2007: 204–206). In the 2005 Sichuan and Anhui Interview Records, the most common description by migrant workers of themselves is *nongmin* (peasants) or *nongcun ren* (village people), sometimes qualified by *dagong* (i.e. *dagong di nongmin*) (working peasants or peasants doing migrant work) (see Chapters 1 and 7). A father of a migrant worker summarizes the connection between his son's identity and eventual return:

> He is a peasant [*nongmin*] after all. Peasants who live and work in the city can be considered half urban and half rural. But fundamentally they are still peasants. Eventually they have to return to the village.
>
> 2005 Sichuan and Anhui Interview Records

A recent survey reports that even among migrants who have lived and worked in the city for a long time and among migrant children who grew up in the city, the peasant identity is dominant, while the sense of belonging to the city is minimal (*Zhongguo qingnianbao* 2007). The steadfastness of the peasant identity is partly because of the hukou divides that exclude peasants from urban resources and benefits, but it is also due to the persistent economic and social disparities between urbanites and peasants. This 42-year-old man who has done migrant work for almost ten years describes the latter factor:

> I am still a peasant. Urban and rural people differ in their standards of living, sources of livelihood, and levels of education. The differences are big. Even if you move to the city and obtain hukou there, you are still not part of the city.
>
> 2005 Sichuan and Anhui Interview Records

The above narratives depict a strong sense among peasant migrants that their place is in the countryside. Meanwhile, the opportunity and desire to obtain the best of both worlds explains why many migrants continue to circulate between the origin and places of migration over a long period of time.

Circulation and migrant labor shortage

Since 2004, many stories of labor shortage in southern China have appeared in both Chinese and non-Chinese media. It is reported, for example, that 1.7 million migrant workers in the Pearl River Delta who went home for the Spring

Festival in 2006 did not return after the holiday, lending support to the speculation that the competitive advantage of China based on cheap, massive peasant labor force is coming to an end (Montlake 2006). This notion, however, is highly disputable. First, as discussed earlier, there is no evidence that mobility in China is declining. Far from it, the rate of migration has continued to increase and is expected to increase into the next decades. Second, agricultural labor surplus, generally estimated in the range of 150 to 200 million, continues to grow (see Chapter 1). Neither agriculture nor rural industrial enterprises are able to absorb such surplus. Rural–urban labor migration will, therefore, remain voluminous in the foreseeable future.

The explanation for labor shortage lies elsewhere, namely, migrants' circularity. Guangdong, including the Pearl River Delta, is known for manufacturing production that thrives on low pay, poor working conditions, and exploitation of migrant workers. As discussed in earlier chapters, peasant migrants' main objective is to boost income and send remittances back home. To them, loyalty, career development, and other objectives that entail long job tenure are unimportant (see Chapter 6). As soon as they learn about jobs with higher pay and better working conditions, they are likely to change jobs, which may involve moving to other parts of the country (Jian and Zhang 2005). In recent years, the Yangtze River Valley has emerged as a competition for migrant workers, luring many of them away from southern China (Xinhua News Agency 2005). A 32-year-old migrant who has worked in Fujian for about ten years comments on factories in Guangdong:

> Migrant labor shortage is due to two reasons. First, the factory pays too little. No one is willing to work there. Second, the factory has a bad reputation. Often, they don't pay you the wages originally agreed on. Or, they use the excuse that you are on probation and pay you less. These things happen a lot in Guangdong. That's why people now are not willing to go there.
>
> 2005 Sichuan and Anhui Interview Records

The competition Guangdong faces is well summarized by this 54-year-old worker: "Now migrants are all going to Shanghai and Fujian. Wages are higher there. The bottom line is, if we can make more money somewhere else, we'll go" (2005 Sichuan and Anhui Interview Records).

The mobility of peasant workers, therefore, has enabled them to seek out better economic and employment opportunities. By "voting with their feet," peasant migrants are also expressing their resistance, albeit in an indirect way, against exploitation (see "Resistance"). Indeed, their mobility has strengthened their collective position to negotiate for better terms. According to a recent report, in 2005 local officials in southern China raised the minimum wage by as much as 34 percent in order to retain migrant workers (Fuller 2005).

The above analysis highlights the following. First, the demand for migrant labor in China remains high. To ensure a guaranteed supply of peasant workers, some companies are considering moving their plants inland where labor is cheaper, and others are retaining migrant workers by paying them more (Fuller 2005). Second, strategies such as splitting the household between two places are all the more important, as they enable migrants to be highly mobile and change jobs and locations of work easily. Third, information (about job opportunities) is traveling faster and peasant migrants have learned the ways to be informed about regional differentials in wages and work-related benefits. They know where to go to find good jobs. Finally, peasant migrants have choices, and their choices are expanding. In previous chapters and earlier this chapter, I have argued that peasant migrants can choose to return to the countryside, and many have done so or plan to do so in the future. Migrant labor shortage shows that peasant workers also have choices between places of work, and many have indeed exercised these choices by following wages to other parts of the country.

Migrants' movements between the origin and destination and among places of work, both underscore the notion that it is circulation, not moving for the purpose of staying, that defines rural–urban labor migration in China. There is no sign that these circulatory movements will decline. Far from it, existing evidence suggests that experienced migrants have already mastered the art of circularity and their children are following their footsteps and joining the flows. This presents tremendous challenge to conventional approaches of studying migration, which rely on definitions of migration and return migration that assume a high degree of permanency and that are not designed to address frequent movements and multiple locations.

Resistance

China's peasant migrant workers are typically portrayed as passive, tolerant, and disorganized. There are neither movements toward unionization nor independent representative organizations for migrant workers. Non-government organizations such as the Migrant Women's Club in Beijing tend to promote social and occupational service rather than labor movement (Jacka 2006). The non-militant nature of the migrant labor regime is, indeed, one of the reasons why investors and employers who prioritize cost minimization are attracted to China.

While peasant migrants as a whole lack representation, their voices are increasingly being heard. The most prominent theme of migrant workers' complaints is companies' delay in paying or outright refusal to pay. The story of Premier Wen Jiabao's meeting with a farmer in October 2003 has been widely publicized in Chinese and non-Chinese media (*Los Angeles Times* 2004). After being told by a woman in Sichuan that her migrant-worker husband had not been paid in a year, Premier Wen intervened, so that within hours the back-wages equivalent to about 2,200 yuan were delivered to her home. The Hu Jintao–Wen Jiabao administration has, indeed, presented itself as one that is more consultative and more sensitive to issues of inequality than previous administrations, a point I shall return to in "Policy."

The same *Los Angeles Times* story cited above also has a picture of migrant workers protesting, on the grounds of back-wages, at a construction site in Beijing. Indeed, labor protests across China have increased in number. Some respondents in the 2005 Sichuan and Anhui Interview Records, likewise, have participated in labor protests. A 35-year-old worker who has worked in Pearl River Delta factories for nine years gives an account of such a protest:

> In the Shenzhen factory, my monthly wage was 800 yuan. But the working hours were long. I worked from 7:30 in the morning to 7:30 in the evening. We had 30 minutes for lunch and 30 minutes for dinner. Overtime was from 7:30 pm to 9:30 pm and sometimes to 10 pm. Overtime pay was only 2.48 yuan per hour, although according to the Shenzhen labor law the minimum overtime pay was 4.30 yuan .... The workers were very unhappy about the long working hours, low wages, and especially low overtime pay. We did not want to work overtime, but we wanted to make more money, so it was a dilemma. Some of us reported the situation to government officials, but nothing happened .... In October 2004, we elected 10 representatives to negotiate with the boss. We demanded that the daily wage increase from 13.2 yuan to 16 yuan and the overtime pay increase from 2.48 yuan to 3.9 yuan per hour. He initially agreed, but the next day he changed his mind. The workers were angry and threatened to strike. The boss then hired some gangsters to beat the 10 representatives up and send them away .... That's when all 6,000 workers of the factory decided to go on strike. We wanted the 10 representatives back. We even took it to the street. The demonstration rally continued on for more than 20 km .... Finally, the city and district governments intervened and guaranteed us that they would take care of the situation .... The Shenzhen Labor Bureau then sent someone to the factory to investigate. They made the boss comply with the law and increase the minimum wage to 480 yuan a month and the minimum overtime pay to 4.30 yuan per hour .... The strike was a success. The workers' conclusion is: without strikes, you make 700 to 800 yuan a month; after a strike, you can make 1,000 yuan or more.
>
> 2005 Sichuan and Anhui Interview Records

The above example illustrates how migrant workers discover the power of organization and protests. Other studies have shown that rural Chinese are, albeit slowly, learning about ways to speak out and take charge (O'Brien and Li 2006: 40). Despite increased awareness of workers' rights, most migrants' protests are small-scale and based on the work place – referred to by Ching Kwan Lee (2007: 192) as "cellular activism" – rather than connected into larger movements. In this light, a protest's chance of success depends much on the willingness of the local government to intervene. This 36-year-old migrant worker describes a failed protest at a Guangdong factory:

> We had no holidays and we worked on Sundays .... We worked overtime practically everyday, until 10 pm and sometimes until midnight. But the

factory did not give us any overtime pay. That was very unfair. But we did not have a choice. The factory's attitude was, "if you don't like it you can leave." One time we had a strike for two days; we demanded an increase in wage. The PSB came and arrested the strike organizer .... So, the strike failed. No improvements in compensation were made .... The government didn't pay attention to the strike. As for the PSB, they only serve the interests of the factory.

2005 Sichuan and Anhui Interview Records: 86

Policy

Since the 1990s, the Chinese government has put forth a series of programs to address various forms of inequality. In 1994, the ambitious "eight–seven program" was launched by the Jiang Zemin–Li Peng administration, aiming at pulling 80 million rural people out of poverty within seven years (Cai 2001: 327).[71] In 1999, President Jiang announced a new program to boost the economic growth of the western region. Dubbed the "Western Development" or "Go West" program, it focuses on reducing the gap between inland areas and the eastern coastal region. During the Sixteenth National Congress of the CCP (November 2002), President Jiang stated that a widening rural–urban gap impedes the progress toward a *xiaokang* society – defined as a society where most of the population are of modest means or middle-class – and that this trend should be reversed (CCP 2002: 19). Since President Hu Jintao and Premier Wen Jiabao came to power in 2003, they have continued to emphasize the importance of helping the have-nots, including the rural poor, migrants, and urban laborers. The Eleventh Five-Year Plan (2006–2010) that was approved in March 2006 highlights the goals of "common prosperity" and "socialist harmonious society," further reinforcing the need to enable disadvantaged groups and less-developed regions to share the fruits of economic growth (Fan 2006).

All of the above represent the central government's responses to the rise in inequality – between the haves and have-nots, between rural and urban people, and between rich and poor regions – that have heightened concern over political and social stability. It is in this context that the Hu–Wen administration addresses issues about peasant migrant workers. Their attention, so far, has been on social protection. For example, the recent Report on the Work of the Government, read by Premier Wen at the Fifth Session of the Tenth National People's Congress on March 5, 2007, draws attention to insurance provision for migrant workers: "We will accelerate the establishment of a social safety net targeted at rural migrant workers in cities, with the focus on signing them up for workers' compensation insurance and medical insurance for major diseases" (*China Daily* 2007b). The Hu–Wen administration's publicized attention toward peasant migrants suggests that the government is listening and is more willing than previous administrations to reach out and listen to voices from below (*China Daily* 2007a; Veeck *et al.* 2007: 123). It is widely reported that a 37-year-old construction worker, who had worked in Beijing for 16 years, was one of a dozen "grassroot personalities" invited to attend a State Council meeting in February 2007 (Xinhua News Agency 2007). Apparently, his ideas about pension and social security coverage for migrant workers were written into the Report on the Work of the Government that was read a month later. Also, extensive consultation took place over several years prior to the drafting and ratification of the Eleventh Five-Year Plan (Fan 2006).

Despite Hu and Wen's open and consultative style of leadership, the solutions offered by the central government for peasant migrants tend to be remedial rather than progressive. The Eleventh Five-Year Plan, for example, sets an expected target to transfer 45 million rural labor into urban sectors over the period 2006 to 2010, which implies that these rural residents can move to cities and towns on a permanent basis (Fan 2006). Yet, the plan includes no specifics on how the hukou system will change to enable such massive shifts. While stating that migrant workers' labor rights should be protected, the plan does not favor an all-out removal of migration restrictions. Rather, it encourages the permanent settlement of rural migrants in medium and small cities and small towns but insists that super-large cities should control their population growth.

The most recent debates on hukou, in fact, center less on whether peasant migrants can obtain urban hukou and more on the actual benefits they can get in cities, with or without urban hukou. In early 2007, a survey of more than 11,000 people revealed that the most important implications of urban hukou are for children to receive education, followed by health and social insurance (*Zhongguo qingnianbao* 2007). In reality, hukou reforms in some cities are carried out in such a manner that urban hukou is disassociated with the above benefits. A recent report indicates that peasant migrants that obtain hukou in Hebi city, Henan, can send their children to local schools but would still not have access to health insurance and other benefits (Liaowan xinwen zhoukan 2007). In Ningxia, many migrants gave up rural hukou and access to farmland for urban hukou in Guyuan city, only to find out that they still are not eligible for benefits that urban residents are entitled to.

Beyond the hukou and permanent migrant paradigms

In Chapter 1, I have argued that the prevailing frameworks that have guided migration research in China – those focusing on abolishing the hukou system as the solution to the plight of peasant migrants, and those assuming that migrants desire to become permanent residents – are inadequate. Obtaining urban hukou alone does not seem to have jump-started peasants' social and economic mobility. At the same time, there is plenty of evidence to show that peasant migrants may not want to stay permanently in the city. Circulation from and to the village home and among places of migrant work, and return migration, have in fact enabled peasants to straddle the city and the countryside and benefit from both. In this book, I have shown that household strategies, the social and power relations that underlie these strategies, and migrants' agency, are fundamental to the story of the Chinese migrant.

I do not wish to suggest, however, that hukou is no longer important. To do that would be throwing the baby out with the bathwater. There is no question that hukou has reinforced rural–urban divides, engendered a two-track migration system, and relegated peasant migrants to the bottom rungs of the city. It is also quite clear that hukou barriers are here to stay for the foreseeable future (Chan and Buckingham 2007; Wang, Fei-Ling 2005). Yet, the roots of Chinese peasants' and peasant migrants' marginality are deep, historical and complex. To significantly enhance their positions would entail steadfast commitment to improving peasants' human capital and the social and economic infrastructure of the countryside on one hand, and to incorporate in meaningful ways peasant migrants into urban society and economy on the other. In this light, the Eleventh Five-Year Plan's calling attention to rural development and the Hu–Wen administration's highlighting social protection of migrant workers, albeit with only vague guidelines on how these could be achieved, are right on the mark (Fan 2006) (see "Policy"). Migration per se may not be at the heart of the problem. In a recent paper, William A.V. Clark (2007) argues that global migration may force us to finally examine and deal with the continuing inequities in the world. Likewise, the gaps – between coastal and inland regions, between the city and countryside, between rural and urban peoples, and between men and women – that persist and are widening in China, may be the real source of the problem.

Unlike 30 years ago, China is functioning more and more like a market economy. Population movements are increasingly reflective of regional disparity and labor demand and supply. While many peasant migrants choose circular migration and eventual return over permanent residence in urban areas, some have managed to leave the countryside for good or are seeking to take roots in the city, as illustrated by Beijing's Zhejiang Village. And, migrants worldwide have shown that their resilience, persistence, and ability to fulfill labor demand in the host economy are more powerful than laws and legislation in shaping human mobility. At the time of this book's writing, the US Congress is debating a new bill that will legalize the nation's estimated 12 million illegal immigrants, aiming at resolving once and for all the dilemma between persistent demand for immigrants' labor and their long-term de facto but illegal residence in the country.[72] By becoming indispensable components of the host country's economy and integrating into the host society and polity, undocumented immigrants in the US are forcing lawmakers to consider compromise. If these migrants' experience can serve as a guide to China's peasant migrants, then the latter's chance of becoming full-fledged urban citizens – if they want to – may very well depend on their long-term value to the urban economy and their incorporation into the political and social fabric of the city.

Notes

1 The International Organization for Migration (2005) estimates that by 2005 there were at least 190 million international migrants – people living outside their country of birth.
2 It took 80 years for the level of urbanization in the United States to increase from about 20 percent to 50 percent, but the same increase is likely to occur over a period of about 30 years in China.
3 The original sample has duplicate records, probably due to processing errors. See Lavely and Mason (2006) for details. After the duplicate records are removed, the sample has a total of 11.5 million observations and is the one used in this book's analyses.
4 The sample has a total of 30,080 observations. For reasons unknown, the total volume of interprovincial migrants based on this sample (31.7 million) is smaller than the volume reported in printed publications (32.3 million) (NBS 2002).
5 The survey was designed in conjunction with Ling Li of Zhongshan University, Kam Wing Chan of University of Washington, and Yunyan Yang of Zhongnan University of Economics and Law as part of a collaborative project.
6 The fieldwork was conducted in collaboration with Ling Li of Zhongshan University.
7 This is part of a collaborative project with Nansheng Bai of Renmin University and other researchers in China.
8 Yunyan Yang notes that there are at least 20 different and related concepts for describing population movements in China (Zhou 2002). Jiao (2002) comments that the definitions of migrants in China are the most complex in the world.
9 Some studies have, however, overlooked this point and compared the two censuses' total migration volumes as if they were based on the same spatial definitions (e.g. Meng 2003).
10 In the US, between the 1990 and 2000 censuses, the rates of intercounty migration declined from 19 percent to 18 percent and the rates of interstate migration declined from 9 percent to 8 percent.
11 The temporal criterion varies greatly between sources and can range from 24 hours to one year (Goodkind and West 2002). Obviously, these variations result in widely varied estimates. Definitions using a short temporal criterion may include transients and travelers (Shen 2002).
12 Combining intercounty and intracounty counts, the 2000 census reports a total of 144.4 million floating population, accounting for 11.6 percent of the nation's population (NBS 2002). This number is consistent with most published sources, which estimate that the floating population was about 30 million in the early 1980s, 70–80 million in the early and mid-1990s, and between 100 million and 140 million in the late 1990s (Bai and Song 2002: 4; Jiao 2002; Solinger 1999a: 18; Wan 2001; Zhong 2000). The 2005 One Percent Population Sample Survey reports further increase of the floating population to 150 million (NBS 2006).

13 The definitions of urban areas and urban population, and hence the measurement of level of urbanization, in China are a constant source of debate. Numerous papers and books have been written on these issues (e.g. Chan 1994; Chan and Xu 1985; Hsu 1994; Ma and Cui 1987; Wu 1994; Zhang and Zhao 1998). Zhou and Ma's (2003) and Chan and Hu's (2003) papers summarize succinctly the changes in definition and the problems associated with each set of definitions used in various Chinese censuses. In the 1982 census, urban population was defined as the total population found within the administrative boundaries of cities and towns. Since the 1980s, however, the criteria for establishing cities and towns have been significantly relaxed and as a result the number of towns and cities increased dramatically. Zhou and Ma, for example, show that the 1982 definition of urban population would have produced a grossly inflated urbanization level of 53.2 percent for 1990. In the 1990 census, a second, more restrictive set of criteria, focusing on population within cities' urban districts and towns' residents' committees, yields a more realistic level of 26.2 percent. In the 2000 census, additional criteria of population density, population size and the extent of built-up areas are used, and an urbanization level of 36.1 percent (November 1) is reported (see also Yu 2001). Most scholars, including Zhou and Ma and Chan and Hu, consider the definition used in the 2000 census more realistic and desirable than those in previous censuses. Yet, since no one set of criteria has been used in two consecutive censuses, levels of urbanization reported in censuses are not comparable. In light of that, Zhou and Ma apply the 2000 census's definition backward to past years and arrive at adjusted year-end urbanization levels of 21.2 percent, 28.4 percent and 36.3 percent, respectively, for 1982, 1990, and 2000.

14 Chan and Hu (2003) estimate that net rural–urban migration (including urban reclassification) accounted for, respectively, 74 percent and 80 percent of China's urban growth during the 1984–1990 period and the 1991–2000 period. These estimates are similar to those of Lu and Wang (2006), who found that rural–urban migration accounted for 79 percent of urban growth during the period 1979 to 2003.

15 There are three types of provincial-level units (Figure 2.1): 22 "provinces" (*sheng*), five "autonomous regions" (*zizhiqu*), and four "centrally administered cities" (*zhixiashi*). The centrally administered cities are also referred to as "centrally administered municipalities," "cities under central administration," "provincial-level cities," "municipal cities," or simply "municipalities." The term "province" is commonly used to refer to all three types of provincial-level units.

16 At the prefecture level, provinces and autonomous regions are subdivided into "prefectures" (*diqu*), "autonomous prefectures" (*zizhizhou*), and "prefecture-level cities" (*dijishi*). Centrally administered cities do not have prefecture-level subdivisions. Since the mid-1990s, a small number of prefecture-level cities, such as Guangzhou and Nanjing, have been upgraded into "secondary provincial-level cities" (*fushengji shi*) (totaling 15 in 2004). However, these secondary provincial-level cities are often categorized as prefecture-level units and thus the basic four-level spatial administration remains intact.

17 Cities refers to county-level cities and the "urban districts" (*shixiaqu*) of prefecture- and provincial-level cities. Counties refers to regular counties (*xian*) and suburban counties (*jiaoxian* or *shixiaxian*) of prefecture- and provincial-level cities.

18 Cities (county-level cities and urban districts) are subdivided into streets, towns and townships, but counties are subdivided into only towns and townships.

19 Mean distance of interprovincial migration shows a slight increase between the 1990 and 2000 census (see Table 5.1).

20 Wu and Treiman (2004) estimate that between the late 1980s and mid-1990s 95 percent of rural residents held agricultural hukou and 90 percent of urban residents held non-agricultural hukou.

21 Work units or *danwei* refers to where urban residents work and more specifically to SOEs, which in the pre-reform period were practically the only employers for urban

residents. Employment in *danwei* also connotes formal employment (e.g. Zhang 2000), as opposed to work that is not associated with *danwei*, such as farming.

22 Many studies have found that rural migrants are marginalized and disadvantaged and that they occupy the lowest social and occupational rungs in urban society (Cai 2002; Fan 2002a; Solinger 1999a). A study of Beijing, Wuxi and Zhuhai, for example, finds that rural–urban migrants are most susceptible to poverty compared with urban residents and urban–urban migrants (Wang and Li 2001). Nevertheless, there are also examples of migrant communities that managed to establish thriving niches in cities, such as the Zhejiang Village in Beijing (Ma and Xiang 1998; Xiang 2005).

23 Commenting on the exploitation peasant migrant workers face, Anita Chan (2001: 9) argues that the hukou system has provided the perfect condition for forced and bonded labor.

24 A variety of terms have been used to describe these fees, the most popular being *jiangshe fei* (development fee), *chengshi jiangshe fei* (urban development fee), *jiangzhen fei* (town development fee), *jiangshe peitao fei* (development and accessory fee) and *zengrong fei* (accommodation fee). The "price" of hukou varies and is largely a function of the administrative status of the city or town. A hukou in prefecture-level cities costs more than that in county-level cities, which in turn is more expensive than that in towns; the fees are higher in the city proper than the outskirts of cities (Cao 2001; Chan and Zhang 1999; Han 1994).

25 In 1994, blue stamp hukou became available in Shanghai, to individuals who invested one million yuan in the city, purchased an apartment 100 square meters or bigger, or were professionals already employed in the city for three years or more. After a specified period of time, holders of blue stamp hukou could apply for a regular Shanghai hukou. In Shenzhen, holders of blue stamp hukou were required to pay the city government 2,000 yuan every year, and upon paying for ten consecutive years would be eligible to apply for regular Shenzhen hukou (Guangdong wailai nongmingong lianhe ketizu 1995; *China Daily* 2001).

26 First, children can now choose to inherit hukou from the father or the mother (previously, hukou was inherited only from the mother). Second, rural persons who have lived in the city for more than one year and whose spouses hold urban hukou may be granted urban hukou. Third, elderly parents whose only children live in cities may be granted urban hukou. Fourth, persons who have made investment, have established enterprises or purchased apartments, who have stable jobs and accommodation, and who have lived more than one year in a city are eligible for urban hukou (Yu 2002: 381).

27 It appears that many cities have not practiced, or have only recently begun to practice, the State Council guidelines. In April 2006, for example, Beijing city government revised its legislation such that children born to fathers who are Beijing residents can obtain Beijing hukou, regardless of their mothers' hukou status (*Beijing chenbao* 2006).

28 The notion of "strictly controlling large cities and some degrees of relaxation for medium and small cities" (*da chengshi yange kongzhi, zhongxiao chengshi shidang fangsong*), which aims at maintaining the urban hierarchy among cities, continues to be popular among policy makers. Thus, urban hukou in large cities, prefecture-level cities, and centrally administered municipalities is still strictly controlled (Yu 2002: 394).

29 In Zhuhai, for example, persons who have lived there for five consecutive years, have a job and a permanent place to stay, have educational attainment above the senior high, have no criminal records and are healthy can apply for local hukou (Cai 2002: 227).

30 In 2001, Beijing government began to implement a hierarchical system of three types (A, B and C) of temporary permits, which determine the services migrants would have access to and the extent of government control over them (Cai 2002: 238). Skilled workers can apply for the most desirable A permit once they arrive.

176 *Notes*

31 This is despite the fact that since August 2001, enterprises in Beijing no longer need to specify hukou requirements in their job advertisements, and owners of private enterprises can apply for Beijing hukou (Cai 2002: 228; *China Daily* 2001).
32 Despite Mao's praises for the Chinese peasants and their historical importance for his rise to power, especially during the Chinese Communist Party–Kuomintang strifes (e.g. Smith 2000: 51), CCP policies continued to favor urbanites.
33 Though recent economic and institutional changes have potential to blur the Rural–urban divide, Rural–urban disparity remains large and has increased over the past two decades (see Chapter 7).
34 The new category "housing change" refers to moving house due to demolition or other reasons. Its inclusion in the 2000 census is probably aimed at reflecting increased residential mobility due to new urban and housing construction in many cities and to the housing reform. From the late 1980s, a housing market began to emerge in many large cities. State-subsidized resources have been made available to facilitate home purchase by state employees, subject to criteria such as the employee's rank and seniority. At the same time, rapid expansion of private housing has boosted residential mobility. Similar types of moves in rural areas are not as prevalent, but large-scale constructions such as the Three Gorges Project necessitate resettlement. "Housing change" is likely to be highly represented by intraurban and intracounty short-distance moves. It represents only 4.5 percent of intercounty migration in the 2000 census (see also Figure 4.2).
35 Liu and Chan (2001) argue that these migration reasons can facilitate the distinction between state-planned migration and self-initiated migration. They caution, however, that the reasons are only rough approximations of actual channels for changing hukou, which are detailed only in the hukou records administered by the Ministry of Public Security.
36 Liu and Chan (2001) use three categories to organize migration reasons: state-sponsored work, self-sponsored work, and family.
37 These ranks mask important differences between men and women, which will be detailed in Chapter 8.
38 The majority of migrants in the 1998 Guangzhou survey are interprovincial migrants, further supporting the observation that distance has not deterred their move to Guangzhou.
39 See Hare (1999b), Wang (2000) and Yang and Guo (1999). Yang and Guo show that marital status and having children are the most deterring factors of female migration in China, Wang observes that most women migrate while they are single, and Hare finds that marital status is an important factor in women's propensity for engaging in wage employment.
40 Although provincial capitals are not always near the geographic centroids of provinces, they are usually among the largest cities in respective provinces and are appropriate proxies for demographic centers of gravity. The average distance of interprovincial moves d is defined as

$$d = \frac{\sum_i \sum_j m_{ij} d_{ij}}{MD} \quad i \neq j$$

where i = origin province
j = destination province
m_{ij} = migration between i and j
d_{ij} = distance between i and j

$$M = \sum_i \sum_j m_{ij} \quad i \neq j$$

$$D = \sum_i \sum_j d_{ij} \quad i \neq j$$

41 Despite the one-child policy, many rural Chinese have managed to have two or more children. In some cases, this is due to local provisions that exempt certain households from the policy. In other cases, villagers are willing to pay the fine for violating the policy. Other means, such as bribery and hiding the pregnancy by moving to another place, are also used to circumvent the birth control restrictions.
42 See also Pang Hui's story in Jacka and Song (2003). When she returned home from migrant work, her husband thought of her money as dirty and refused to let her stay.
43 Skinner (1976) draws a distinction between residence in one's native place and abode in one's place of work. He argues that residence is permanent and can be interpreted as an ascribed characteristic whereas abode is impermanent and can vary.
44 Tyson and Tyson (1996) use the term *mangliu*, which is most often used to describe disorderly migrants that entered cities during the 1950s (see Chapter 2), to refer to rural migrant workers in the 1990s.
45 *Tongxiang* or *laoxiang* refers to people from the same native place. Skinner's (1976) study of late imperial China highlights *tongxiang* as links between migrants and their native place that ease the process of migration.
46 Knight and Song (1999: 319–338) document that rural–urban income inequality increased between 1988 and 1995 and that it is a major contributor to overall income inequality in China. Li (2003) observes that rural–urban inequality accounted for 40 percent of income inequality in 1988 and that the absolute level of rural–urban inequality continued to increase between 1988 and 1995. Knight and Song (1999: 338) note that China's rural–urban income gap is only surpassed by South Africa and Zimbabwe. Li (2003) shows that if urban non-monetary income such as material income and subsidies are taken into account, then in 2000 the ratio of urban income to rural income in China was 3.62:1 and was the largest in the world. Similarly, Sicular *et al.* (2007) note that the gap in per capita income between rural and urban areas widened during the reform period, reaching a ratio of three to one.
47 Cai (2001: 344) cites a survey which estimates that the 41.4 million rural migrants in 1994 had a total migrant income of 151.1 billion yuan, averaging 3,676 yuan per migrant.
48 Every year RSEST of NBS conducts a sample survey of rural households across China (see Chapter 1). In both 2002 and 2003, 68,000 households were surveyed.
49 Other studies on specific origins of migrants also support the notion that migrant work is a major factor of income growth in rural areas (Du and Bai 1997: 130; Qiu *et al.* 2004).
50 Cai (2000: 198) estimates that average remittance by rural–urban migrants is about 2,000 yuan per year.
51 The survey was conducted by the Institute of Economics, CASS in collaboration with NBS. It included 19 provinces and 7,998 rural households. Of all the households surveyed, 1,181 received remittances from migrants that averaged 2,190 yuan.
52 Lian (2002) estimates that in 1995 the more than five million labor migrants from Sichuan sent back remittances totaling 15.7 billion yuan, averaging about 3,140 yuan per migrant. The remittances amounted to about one-tenth of Sichuan's total agricultural output (152 billion) that year (Sichuan Statistical Bureau 1996: 21).
53 The survey was conducted in the fall of 1995 in Xiayi county of Henan province, including 309 randomly selected households from three townships (Hare 1999).
54 Lower and higher estimates have also been reported (Han *et al.* 2002). Qiu (2001) estimates that an additional 13 million agricultural surplus labor would be created between 2001 and 2010.

178 *Notes*

55 For example, Yang (2003) estimates that the per capita net income of migrant households is higher than that of non-migrant households by 1,050 yuan.
56 Due partly to poverty, many rural Chinese do not change clothes often. Many villagers I met during the Gaozhou field study wore practically the same clothes every day and seldom changed clothes.
57 *Putonghua* is standard Mandarin that is considered a more formal and sophisticated medium of conversation, one that is especially associated with urban people, than a wide range of dialects spoken across the countryside.
58 This expression is commonly used to describe rural people whose wearing of colorful clothes is interpreted as lack of sophistication and taste.
59 Marriage migration refers to migration that occurs at the beginning of marriage. Migration as a family and moves for family reunification are, strictly speaking, not marriage migration. In the terminology of Chinese censuses, these types of moves are usually considered "joining family" migration (see Table 4.1), defined as migration to (re)join the spouse after prior geographical separation due to job-related and other reasons.
60 A survey in Beijing finds that less than one percent of young women are willing to marry men with income lower than theirs (Ye 1997).
61 Traditionally, especially in rural areas, men are expected to marry in their 20s and certainly by the age of 30. Men who are not married by the age of 30 are considered "above marriageable ages." Women are expected to get married before they reach their mid-20s (e.g. Tan 1996a).
62 Sun's (1991: 39–42) book describes several anecdotal examples of marriage migration that eventually resulted in divorce. An older man married an Inner Mongolian woman 18 years younger than he, but later divorced her because of difficulties in dealing with the age difference. Another man was introduced by a marriage broker to a woman from a poor province, married her, but upon finding out that she had cheated about her age and past marriage, divorced her. One woman from Anhui, a poor province, introduced herself to an older man who was eager to find a marriage partner, and then took his money and left only a few days after the wedding.
63 As discussed in Chapter 4, migrants that select "seeking work in industry and business" as their migration reason are primarily labor migrants.
64 Huang (2002) finds that men in Jiangsu that marry women from other provinces tend to be older and have relatively low income.
65 Gaozhou county was upgraded to be a county-level city in 1995. But because it is primarily an agricultural county, with only a small number of industrial enterprises, in this book I continue to refer to it as a county. As of 1996, Gaozhou county comprised 27 townships and a total population of 1.4 million (GDSB 1997).
66 Pseudonyms are used for interviewees from the 1999 Gaozhou field study.
67 According to an informant, brides from neighboring countries, especially Vietnam, have also been imported via unofficial routes to marry men that have difficulties finding mates.
68 Multi-generation households living under one roof are no longer the norm in rural China. Rather, sons are expected to move out and establish new households upon marriage. Among the 76 households surveyed, there are six one-generation, 57 two-generation and only 13 three-generation households.
69 In response to reports that one million migrants will be expelled from the city for the duration of the Olympics, Beijing officials denied that such plans exist and indicated that other measures would be taken to maintain public order and manage traffic during the Games (Xinhua News Agency 2006).
70 Some surveys in Shanghai and Beijing find that significant proportions of migrants came as families (see Peng *et al.* 1998 and Song and Gu 1999).
71 This ambitious goal was publicized to the world in a speech delivered in 1995 in Copenhagen by former Premier Li Peng. Recent government announcements describe

the program as completed and its goals basically achieved (CCP 2002: 4). For example, in a speech by Gao Hongbin, head of the State Council's Poverty Alleviation Office, on June 19, 2001, he announced that by the end of 2000 the rural population in poverty had declined to 30 million (Wuguo 2001). A document by the State Council in early 2004 emphasizes a vision of *xiaokang* society instead (CASS 2004: 1).
72 The proposed bill also includes a temporary-worker program and steps to strengthen border security and workplace enforcement.

References

Adams Jr, Richard H. and Page, John (2003) "International migration, remittances and poverty in developing countries," *World Bank Policy Research Working Paper*, 3179.

Alexander, Peter and Chan, Anita (2004) "Does China have an apartheid pass system?" *Journal of Ethnic and Migration Studies*, 30(4): 609–629.

Amin, Ash (1999) "An institutionalist perspective on regional economic development," *International Journal of Urban and Regional Research*, 23(2): 365–378.

Bader, Jeffrey A. (2006) "China's rise: what it means for the rest of us," speech, [www.brookings.edu/views/speeches/bader/20060907.htm], September 7 (accessed May 18, 2007).

Bai, Nansheng and Song, Hongyun (eds) (2002) *Huixiang, Hai Shi Jincheng: Zhongguo Nongcun Waichu Laodongli Huiliu Yanjiu (Return to the Village or Enter the City: Research on the Return Migration of Rural Migrant Labor in China)*, Beijing: Zhongguo caizheng jingji chubanshe (in Chinese).

Ban, Maosheng and Zhu, Chengsheng (2000) "*Huji gaige di yanjiu zhuangkuang ji shiji jinzha*" (Status of the studies on household registration system reform and the progress), *Renkou Yu Jingji (Population and Economics)*, 2000(1): 46–51 (in Chinese).

Banister, Judith (2004) "Shortage of girls in China today," *Journal of Population Research*, 21(1): 19–45.

Banister, Judith and Taylor, Jeffrey R. (1989) "China: surplus labour and migration," *Asia-Pacific Population Journal*, 4(4): 3–20.

Banister, Judith and Harbaugh, Christina Wu (1992) "Rural labor force trends in China," *Agricultural and Trade Report* (Situation and Outlook Series, Economic Research Service, United States Department of Agriculture).

Bauer, John, Wang, Feng, Riley, Nancy E. and Zhao, Xiaohua (1992) "Gender inequality in urban China," *Modern China*, 18(3): 333–370.

BBC News (2005) "China rethinks peasant 'apartheid,'" [http://news.bbc.co.uk/2/hi/asia-pacific/4424944.stm], November 10 (accessed June 16, 2006).

Beijing chenbao (Beijing Morning Post) (2006) "*Beijing shi zhengfu xiugai fagui mingque xingshenger suifu ruhu zhengce*" (Beijing city government revises legislation to confirm the policy of new-born children following the hukou of fathers), [www.china.com.cn/chinese/kuaixun/1176839.htm], April 7 (accessed May 8, 2007) (in Chinese).

Beijing sheke guihu (Beijing Social Science Planning) (2000) "*Zhongguo liudong renkou meinian jiang zengjia wubaiwan*" (China's floating population increases by five million every year), [www.bipopss.gov.cn/bipssweb/show.aspx?id=15056&cid=51], October 20 (accessed May 8, 2007) (in Chinese).

Bian, Yanjie, Logan, John R., and Shu, Xiaoling (2000) "Wage and job inequalities in the working lives of men and women in Tianjin," in Barbara Entwisle and Gail Henderson (eds), *Re-Drawing Boundaries: Work, Households, and Gender in China*, Berkeley, CA: University of California Press, 111–133.

Boehm, Thomas P., Herzog, Henry W., and Schlottmann, Alan M. (1991) "Intra-urban mobility, migration, and tenure choice," *Review of Economics and Statistics*, 73(1): 59–68.

Bonney, Norman and Love, John (1991) "Gender and migration: Geographical mobility and the wife's sacrifice," *Sociological Review*, 39(2): 335–348.

Borjas, George J. (1989) "Immigrant and emigrant earnings: A longitudinal study," *Economic Inquiry*, 27(1): 21–37.

Bossen, Laurel (1994) "*Zhongguo nongcun funu: shime yuanyin shi tamen liuzai nongtianli?*" (Chinese peasant women: What caused them to stay in the field?), in Xiaojiang Li, Hong Zhu and Xiuyu Dong (eds), Xingbie Yu Zhongguo (*Gender and China*). Beijing: Sanlian Shudian, 128–154 (in Chinese).

Breman, J. (1976) "A dualistic labour system? A critique of the informal sector concept," *Economic and Political Weekly*, 11(48): 1870–5.

Brown, L. and Moore, E. (1970) "The intra-urban migration process: a perspective," *Geografisker Annaler*, 52: 1–13.

Buckley, Cynthia (1995) "The myth of managed migration: Migration control and market in the Soviet period," *Slavic Review*, 54(4): 896–914.

Cai, Fang (1997) "*Qianyi juece zhong de jiating jiaose he xingbie tezhen*" (The role of family and gender characteristics in migration decision making), *Renkou Yanjiu* (*Population Research*), 21(2): 7–21 (in Chinese).

—— (2000) *Zhongguo Liudong Renkou Wenti* (*China's Floating Population*), Zhengzhou, China: Henan renmin chubanshe (in Chinese).

—— (2001) *Zhongguo Renkou Liudong Fangshi Yu Tujing (1990–1999 Nian)* (*The Means and Paths of Population Migration in China (1990–1999)*), Beijing: Shehui kexue wenxian chubanshe (Social Science Documentation Publishing House) (in Chinese).

—— (2002) *Zhongguo Renkou Yu Laodong Wenti Baogao: Chengxiang Jiuye Wenti Yu Duice* (*Report on China's Population and Labor: Employment Issues and Strategies in Urban and Rural Areas*), Beijing: Shehui kexue wenxian chubanshe (Social Sciences Documentation Publishing House) (in Chinese).

—— (2003) *Zhongguo Renkou Yu Laodong Wenti Baogao: Zhuangui Zhong Di Chengshi Pinkun Wenti* (*Report on China's Population and Labor: Urban Poverty in Transitional China*), Beijing: Shehui kexue wenxian chubanshe (Social Sciences Documentation Publishing House) (in Chinese).

Cai, Fang and Wang, Dewan (2003) "Migration as marketization: What can we learn from China's 2000 census data?" *China Review*, 3(2): 73–93.

Calavita, Kitty (1992) *Inside the State: The Bracero Program, Immigration, and the I.N.S.*, New York: Routledge.

Cao, Jing-chun (2001) "*Guanyu lanyin hukou wenti di sikao*" (Some thoughts on the blue stamp hukou), *Population and Economics*, 2001(6): 15–21, 66 (in Chinese).

Cao, Xiangjun (1995) "*Zhongguo nongcun laodongli liudong yu renkou qianyi yanjiu zongshu*" (Summary of research on rural labor flows and population migration in China), *Nongcun Jingji Yanjiu Cankao* (*Rural Agricultural Research*), 1995(2): 23–33 (in Chinese).

Cartier, Carolyn (1998) "Women and gender in contemporary China," in Carolyn V. Prorok and Kiran Banga Chhokar (eds), *Asian Women and Their Work: A Geography*

of Gender and Development, Indiana, PA: National Council on Geographic Education, 1–7.
CASS (Chinese Academy of Social Sciences) and National Bureau of Statistics (2003) *2002–2003 Nian Zhongguo Nongcui Jingji Xingshi Fenxi Yu Yuce* (*Analysis and Forecast on China's Rural Economy 2002–2003*), Beijing: Shehui kexue wenxian chubanshe (Social Sciences Documentation Publishing House) (in Chinese).
—— (2004) *2003–2004 Zhongguo Nongcui Jingji Xingshi Fenxi Yu Yuce* (*Analysis and Forecast on China's Rural Economy 2003–2004*), Beijing: Shehui kexue wenxian chubanshe (Social Sciences Documentation Publishing House) (in Chinese).
CCP (Chinese Communist Party) (2002) *Zhongguo Gongchandang Di Shiliu Ci Quanguo Daibiao Dahui Wenjian Huibian* (*Collection of Documents From the CCP's Sixteenth National Congress*), Beijing: Renmin chubanshe (in Chinese).
Cerrutti, Marcela and Massey, Douglas S. (2001) "On the auspices of female migration from Mexico to the United States," *Demography*, 38(2): 187–200.
Chan, Anita (2001) *China's Workers Under Assault: The Exploitation of Labor in a Globalizing Economy*, Armonk, NY: M.E. Sharpe.
Chan, Kam Wing (1994) *Cities With Invisible Walls: Reinterpreting Urbanization in Post-1949 China*, Hong Kong: Oxford University Press.
—— (1996) "Post-Mao China: a two-class urban society in the making," *International Journal of Urban and Regional Research*, 20(1): 134–150.
—— (2003) "Chinese census 2000: new opportunities and challenges," *China Review*, 3(2): 1–12.
Chan, Kam Wing and Xu, Xueqiang (1985) "Urban population growth and urbanization in China since 1949: Reconstructing a baseline," *China Quarterly*, 104: 583–613.
Chan, Kam Wing and Zhang, Li (1999) "The hukou system and rural–urban migration in China: Processes and changes," *China Quarterly*, 160: 818–855.
Chan, Kam Wing and Hu, Ying (2003) "Urbanization in China in the 1990s: New definition, different series, and revised trends," *China Review*, 3(2): 48–71.
Chan, Kam Wing and Buckingham, Will (2007) "Is China abolishing the hukou system?" *China Quarterly*, forthcoming.
Chan, Kam Wing, Liu, Ta, and Yang, Yunyan (1999) "Hukou and Non-hukou migrations in China: comparisons and contrasts," *International Journal of Population Geography*, 5: 425–48.
Chant, Sylvia (1996) "Women's roles in recession and economic restructuring in Mexico and the Philippines," *Geoforum*, 27(3): 297–327.
Chant, Sylvia and Radcliffe, Sarah (1992) "Migration and development: The importance of gender," in Sylvia Chant (ed.), *Gender and Migration in Developing Countries*, New York: Belhaven Press, 1–29.
Chen, Yangle (2001) "The studies of scale of the Chinese surplus agricultural labor force and the economic price of its being detained," *Renkou Yu Jungji* (*Population and Economics*), 2001(2): 52–58.
Chen, Yintao (1999) "Marriage attitudes and problems among migrant women workers: Survey results in Guangdong," in Nora Chiang and Cathy Sung (eds), *Population, Urban and Regional Development in China*, Taipei, Taiwan: China Studies Program, Population Studies Center, National Taiwan University, 181–192.
Cheng, Lucie and Hsiung, Ping-Chun (1992) "Women, export-oriented growth, and the state: The case of Taiwan," in Richard P. Appelbaum and Jeffrey Henderson (eds), *States and Development in the Asian Pacific Rim*, Newbury Park, CA: Sage Publications, 233–266.

Cheng, Teijun and Selden, Mark (1994) "The origins and social consequences of China's hukou system," *China Quarterly*, 139: 644–668.

Cheng, Youqi (1994) "*Zhongguo duiwai maoyi di quyu fanzhan zhanlue yanjiu*" (The regional development strategy of China's foreign trade), *Jinji Dili* (*Economic Geography*), 14(2): 75–80 (in Chinese).

Chiang, Nora (1999) "Research on the floating population in China: Female migrant workers in Guangdong's township–village enterprises," in Nora Chiang and Cathy Song (eds), *Population, Urban and Regional Development in China*, Taipei, Taiwan: Population Studies Center, National Taiwan University, 163–180.

China Daily (2001) "China reforms domicile system," [http://service.china.org.cn/link/wcm/Show_Text?info_id=19022&p_qry=2000%20and%20census], September 12 (accessed July 18, 2003).

—— (2004) "Hukou system must reform step by step," [http://en-1.ce.cn/Business/Macro-economic/200409/16/t20040916_1772934.shtml], September 16 (accessed June 6, 2005).

—— (2005) "Discrimination in job market," [www.chinadaily.com.cn/english/doc/2005-05/09/content_440368.htm], May 9 (accessed June 3, 2005).

—— (2006) "Experts: Improve welfare of 200 million rural workers," [www.chinadaily.com.cn/china/2006-05/01/content_581684.htm#], May 1 (accessed May 11, 2007).

—— (2007a) "Premier Wen celebrates new year with people," [http://en.beijing2008.cn/88/10/article214021088.shtml], February 20 (accessed April 12, 2007).

—— (2007b) "Report on the work of the government," [www.chinadaily.com.cn/china/2007-03/17/content_830171.htm], March 17 (accessed April 13, 2007).

—— (2007c) "WTO: China overtakes US as second biggest exporter," www.chinadaily.com.cn/china/2007-04/12/content_849420.htm], April 12 (accessed May 18, 2007).

Christiansen, Flemming (1992) "'Market transition' in China: The case of the Jiangsu labor market, 1978–1990," *Modern China*, 18(1): 72–93.

Clark, William A. V. (2007) "Human mobility in a globalizing world: Urban development trends and policy implications," in H. S. Geyer (ed.), *International Handbook of Urban Policy*, Vol I: *Contentious Global Issues*, Edward Elgar.

Clark, William A. V. and Huang, Youqin (2006) "Balancing move and work: Women's labour market exits and entries after family migration," *Population, Space and Place*, 12: 31–44.

Clark, William A. V., Duerloo, M. C., and Dieleman, F. M. (1984) "Housing consumption and residential mobility," *Annals of the Association of American Geographers*, 74: 29–43.

Congressional-Executive Commission on China (2005) "Recent Chinese hukou reforms," [www.cecc.gov/pages/virtualAcad/Residency/hreform.php] (accessed June 16, 2006).

Cook, Sarah and Maurer-Fazio, Margaret (1999) "Introduction to the special issue on the Workers' State Meets the Market: Labour in China's Transition," *Journal of Development Studies*, 35(3): 1–15.

Croll, Elisabeth (1981) *The Politics of Marriage in Contemporary China*, Cambridge: Cambridge University Press.

—— (1984) "The exchange of women and property: Marriage in post-revolutionary China," in Renee Hirschon (ed.), *Women and Property – Women As Property*, London: Croom Helm, 44–61.

Croll, Elisabeth J. and Huang, Ping (1997) "Migration for and against agriculture in eight Chinese villages," *China Quarterly*, 149: 128–146.

Da jiyuan (Big Century) (2006) "*Zhongguo liudong renkou da 1.5 yi ren yu 8 cheng wen nongmingong*" (China's floating population reaches 150 million and 80 percent are rural migrant workers), [www.epochtimes.com/gb/6-10/30/n1503150.htm], October 30 (accessed May 8, 2007) (in Chinese).

Davin, Delia (1997) "Migration, women and gender issues in contemporary China," in T. Sharping (ed.), *Floating Population and Migration in China*, Hamburg: Institut für Asienkunde, 297–314.

—— (1998) "Gender and migration in China," in Flemming Christiansen and Junzuo Zhang (eds), *Village Inc.: Chinese Rural Society in the 1990s*, Surrey, UK: Curzon Press, 230–240.

—— (1999) *Internal Migration in Contemporary China*, London: MacMillan Press.

Davis, Deborah (1999) "Self-employment in Shanghai: A research note," *China Quarterly*, 157: 22–43.

De Jong, Gordon F. (2000) "Expectations, gender, and norms in migration decision-making," *Population Studies*, 54(3): 307–319.

Deshingkar, Priya (2005) "The role of circular migration in economic growth," *Entwicklung & Landlicher Raum*, [www.rural-development.de/fileadmin/rural-development/volltexte/2005–05/ELR_dt_10-12.pdf] (accessed November 6, 2006).

Ding, Jinhong (1994) "*Zhongguo renkou shengji xianyi de yuanyin bie liuchang tezhen taixi*" (Characteristics of cause-specific rates of inter-provincial migration in China), *Renkou Yanjiu* (*Population Research*), 18(1): 14–21 (in Chinese).

Ding, Jinhong, Liu, Zhenyu, Cheng, Danming, Liu, Jin, and Zou, Jianping (2005) "*Zhongguo renkou qignyi di quyu chayi yu liuchang tezheng*" (Regional differences and flow characteristics of migration in China), *Dili Xuebao* (*Acta Geographica Sinica*), 60(1): 106–114 (in Chinese).

Domanski, Henryk (1990) "Dynamics of labor market segmentation in Poland, 1982–1987," *Social Forces*, 69(2): 423–438.

Du, Ying (2000) "Rural labor migration in contemporary China: an analysis of its features and the macro context," in Loraine A. West and Yaohui Zhao (eds), *Rural Labor Flows in China*, Berkeley, CA: Institute of East Asian Studies, 67–100.

Du, Ying and Bai, Nansheng (eds) (1997) *Zouchu Xiangcun* (*Leaving the Village*), Beijing: Jingji kexue chubanshe (Economic Science Press) (in Chinese).

Duan, Chengrong (2003) "*Lun liudong renkou de shehui shiying: jiantan Beijingshi liudong renkou wenti*" (On the social adaptability of the floating population), *Yunnan Daxue Xuebao* (*Shehui Kexueban*) (*Journal of Yunnan University* (*Social Science*)), 2(3): 54–60 (in Chinese).

Duan, Chengrong and Sun, Yujing (2006) "*Woguo liudong renkou tongji koujing di lishi biandong*" (Changes in the scope and definition of the floating population in China's censuses and surveys), *Renkou Yanjiu* (*Population Research*), 30(4): 70–76 (in Chinese).

Duan, Jianping (1998) "*Nongcun waichu wugong qingnian di shequ dingwei*" (Social positions of young peasant migrants), in Zhongshan University Anthropology Department (ed.), *Waichu Wugong Diaocha Yanjiu Wenji* (*Collection of Papers on Surveys of Peasant Out-Migration*), Guangzhou, China: Zhongshan University Anthropology Department, 74–81 (in Chinese).

Eastday.com (2002) "Blue-cover residency ends," [http://service.china.org.cn/link/wcm/Show_Text?info_id=29977&p_qry=2000%20and%20census], April 2 (accessed July 18, 2003).

Ebrey, P. (1991) "Introduction," in R. Watson and P. Ebrey (eds), *Marriage and Inequality in Chinese Society*, Berkeley: University of California Press.
Eder, James F. (2006) "Gender relations and household economic planning in the rural Philippines," *Journal of Southeast Asian Studies*, 37(3): 397–413.
Editorial group for *guomin jingi he shehui fazhan di shiyi ge wunian guihua gangyao xuexi fudao* (2006) *Guomin Jingi He Shehui Fazhan Di Shiyi Ge Wunian Guihua Gangyao Xuexi Fudao* (*Summary of the Eleventh Five-Year Plan for National Economic and Social Development: Study Guide*), Beijing: CCP Party School (in Chinese).
Entwisle, Barbara and Henderson, Gail E. (2000) "Conclusion: re-drawing boundaries," in Barbara Entwisle and Gail E. Henderson (eds), *Re-Drawing Boundaries: Work, Households, and Gender in China*, Berkeley, CA: University of California Press, 295–303.
Fan, C. Cindy (1995) "Of belts and ladders: state policy and uneven regional development in post-Mao China," *Annals of the Association of American Geographers*, 85(3): 421–449.
—— (1996) "Economic opportunities and internal migration: A case study of Guangdong Province, China," *Professional Geographer*, 48(1): 28–45.
—— (1999) "Migration in a socialist transitional economy: Heterogeneity, socioeconomic and spatial characteristics of migrants in China and Guangdong province," *International Migration Review*, 33(4): 950–983.
—— (2000) "Migration and gender in China," in Chung Ming Lau and Jianfa Shen (ed.), *China Review 2000*, Hong Kong: Chinese University Press, 217–248.
—— (2001) "Migration and labor market returns in urban China: Results from a recent survey in Guangzhou," *Environment and Planning A*, 33(3): 479–508.
—— (2002a) "The elite, the natives, and the outsiders: Migration and labor market segmentation in urban China," *Annals of the Association of American Geographers*, 92(1): 103–124.
—— (2002b) "Permanent migrants, temporary migrants, and the labour market in Chinese cities," in Kwan-yiu Wong and Jianfa Shen (eds), *Resource Management, Urbanization and Governance in Hong Kong and the Zhujiang Delta*, Hong Kong: Chinese University Press, 55–78.
—— (2003) "Rural–urban migration and gender division of labor in China," *International Journal of Urban and Regional Research*, 27(1): 24–47.
—— (2004a) "Gender differences in Chinese migration," in Chiao-min Hsieh and Max Lu (eds), *Changing China: A Geographic Appraisal*, Boulder, CO: Westview Press, 243–268.
—— (2004b) "Out to the city and back to the village: The experiences and contributions of rural women migrating from Sichuan and Anhui," in Arianne M. Gaetano and Tamara Jacka (eds), *On the Move: Women in Rural-to-Urban Migration in Contemporary China*, New York: Columbia University Press, 177–206.
—— (2004c) "The state, the migrant labor regime, and maiden workers in China," *Political Geography*, 23(3): 283–305.
—— (2005a) "Interprovincial migration, population redistribution, and regional development in China: 1990 and 2000 census," *Professional Geographer*, 57(2): 295–311.
—— (2005b) "Modeling interprovincial migration in China, 1985–2000," *Eurasian Geography and Economics*, 46(3): 165–184.
—— (2006) "China's Eleventh Five-Year Plan (2006–2010): From 'getting rich first' to 'common prosperity,'" *Eurasian Geography and Economics*, 47(6): 708–723.
Fan, C. Cindy and Huang, Youqin (1998) "Waves of rural brides: Female marriage migration in China," *Annals of the Association of American Geographers*, 88(2): 227–251.

Fan, C. Cindy and Li, Ling (2002) "Marriage and migration in transitional China: A field study of Gaozhou, western Guangdong," *Environment and Planning A*, 34(4): 619–638.

Fan, Lida (1997) *"Renkou qianyi dui pinkun diqu fazhan di yingxiang"* (Influences of migration on the development of poor areas), *Renkou Xueken* (*Population Journal*), 1997(5): 29–33 (in Chinese).

Farer, Tom J. (1995) "How the international system copes with involuntary migration: Norms, institutions and state practice," *Human Rights Quarterly*, 17(1): 72–100.

Fazhi wanbao (*Legal System Evening News*) (2006) "Zhaopinhui zhi ren Beijing hukou, yi zhanwei ban xiaoshi ju 10 ren" (Job fair recognizes only Beijing hukou; 10 people rejected from one booth), [http://beijing.qianlong.com/3825-2006/07/19/2540@3315766.htm], July 19 (in Chinese) (accessed February 20, 2007).

Fincher, Ruth (1993) "Gender relation and the geography of migration (commentary)," *Environment and Planning A*, 25: 1703–1705.

—— (2007) "Space, gender and institutions in processes creating difference," *Gender, Place and Culture*, 14(1): 5–27.

Fuller, Thomas (2005) "China feels a labor pinch," *International Herald Tribune*, [www.iht.com/articles/2005–04/20/news/costs.php], April 20 (accessed May 7, 2007).

Gaetano, Arianne M. (2004) "Filial daughters, modern women: Migrant domestic workers in post-Mao Beijing," in Arianne M. Gaetano and Tamara Jacka (eds), *On the Move: Women in Rural-to-Urban Migration in Contemporary China*, New York: Columbia University Press, 41–79.

Gaetano, Arianne M. and Jacka, Tamara (eds) (2004) *On the Move: Women in Rural-to-Urban Migration in Contemporary China*, Columbia University Press.

GDPPCO (Guangdong Province Population Census Office) (1992) *Guangdong Sheng 1990 Nien Renkou Pucha Ziliao* (*Tabulations on the 1990 Population Census of Guangdong Province*), Vol I, Taishan: Zhongguo Tongji Chubanshe (in Chinese).

GDSB (Guangdong Statistical Bureau) (1997) *Guangdong Tongji Nianjian* (*Statistical Yearbook of Guangdong*) *1997*, Beijing: Zhongguo Tongji Chubanshe (in Chinese).

Gmelch, George (1980) "Return migration," *Annual Review of Anthropology*, 9: 135–159.

Goldscheider, Calvin (1987) "Migration and social structure: Analytic issues and comparative perspectives in developing nations," *Sociological Forum*, 2(4): 674–696.

Goldstein, Alice and Guo, Shenyang (1992) "Temporary migration in Shanghai and Beijing," *Studies in Comparative International Development*, 27(2): 39–56.

Goldstein, Alice, Guo, Zhigang, and Goldstein, Sidney (1997) "The relation of migration to changing household headship patterns in China, 1982–1987," *Population Studies*, 51: 75–84.

Goldstein, Sidney and Goldstein, Alice (1991) *Permanent and Temporary Migration Differentials in China*, Honolulu: East-West Population Institute, Papers of the East-West Population Institute, No. 117.

Goldstein, Sidney, Liang, Zai, and Goldstein, Alice (2000) "Migration, gender, and labor force in Hubei province, 1985–1990," in Barbara Entwisle and Gail Henderson (eds), *Re-Drawing Boundaries: Work, Households, and Gender in China*, Berkeley, CA: University of California Press, 214–230.

Gonganbu huzheng guanliju (Ministry of Public Security) (1997) *Quanguo Zanzhu Renkou Tongji Ziliao Huibian* (*Tabulations of the National Temporary Population Statistics*), Beijing: Zhongguo renmin gongan daxue chubanshe (in Chinese).

Goodkind, Daniel and West, Loraine A. (2002) "China's floating population: Definitions, data and recent findings," *Urban Studies*, 39(12): 2237–2250.

Goodman, John L. (1981) "Information, uncertainty, and the microeconomic model of migration decision making," in Gordon F. De Jong and Robert W. Gardner (eds), *Migration Decision-Making: Multidisciplinary Approaches to Microlevel Studies in Developed and Developing Countries*, New York: Pergamon Press, 130–148.

Gu, Shengzu (1992) "*Zhongguo lianglei renkou qianyi bijiao yanjiu*" (Comparison of two types of migration in China), *Zhongguo Renkou Kexue (Chinese Journal of Population Science)*, 4(1): 75–84 (in Chinese).

Gu, Shengzu and Jian, Xinhua (eds) (1994) *Dangdai Zhongguo Renkou Liudong Yu Chengzhenhua (Population Movement and Urbanization in Contemporary China)*, Wuhan, China: Wuhan daxue chubanshe (Wuhan University Press) (in Chinese).

Guangdong wailai nongmingong lianhe ketizu (Joint research group on Guangdong's labor migrants) (1995) "*Jingying yimin yu xinxing da chengshi zhanlue*" (Elite migration and the strategies of newly developed large cities), *Nongcun Jingji Yanjiu Cankao (Rural Economic Research)*, 1995(2): 1–15 (in Chinese).

Guangming ribao (Guangdong Daily) (2006) "*Zhou Yixing: Chengzhenhua bushi yuekua yuehao*" (Zhou Yixing: Rapid urbanization is not necessarily good), March 27 (in Chinese) (accessed December 11, 2006).

Han, Jijiang, Wang, Jiachuan, and Shi, Jianmin (2002) "*Nongcun laodongli jiuye wenti shi nongcun jingji huode tupo di guanjian: guonei nongcun laodongli jiiyu wenti yanjiu zongshu*" (The issue of rural labor employment is the key of rural economy development: a summary of domestic studies on rural labor employment), *Renkou Yu Jingji (Population and Economics)*, 2002(2): 62–65 (in Chinese).

Han, Jun (1994) "Breaking the rural–urban separation," *Shidian (Viewpoint)*, June 8: 6.

Hao, Hongsheng, Du, Peng, Lin, Fude, Liu, Shuang, Sung, Jian, Chen, Gong, Chen, Yi, and Liu, Wenhai (1998) "*Wuguo da chengshi wailai renkou guanli wenti yu duice*" (Non-native population in China's large cities: Management measures), *Renkou Yanjiu (Population Research)*, 22(1 and 2): 13–20 and 28–35 (in Chinese).

Hare, Denise (1999a) "'Push' versus 'pull' factors in migration outflows and returns: Determinants of migration status and spell duration among China's rural population," *Journal of Development Studies*, 35(3): 45–72.

—— (1999b) "Women's economic status in rural China: household contributions to male–female disparities in the wage labor market," *World Development*, 27(6): 1011–1029.

Hare, Denise and Zhao, Shukai (2000) "Labor migration as a rural development strategy: A view from the migration origin," in *Rural Labor Flows in China*, Berkeley, California: Institute of East Asian Studies.

Harrell, Stevan (2000) "The changing meanings of work in China," in Barbara Entwisle and Gail Henderson (eds), *Re-Drawing Boundaries: Work, Households, and Gender in China*, Berkeley, CA: University of California Press, 67–78.

Harris, J. R. and Todaro, M. P. (1970) "Migration, unemployment and development: A theoretical analysis," *American Economic Review*, 60: 126–142.

He, Canfei and Gober, Patricia (2003) "Gendering interprovincial migration in China," *International Migration Review*, 37(4): 1220–1251.

He, Jiaosheng and Pooler, Jim (2002) "The regional concentration of China's interprovincial migration flows, 1982–90," *Population and Environment*, 24(2): 149–182.

Hershatter, Gail (2000) "Local meanings of gender and work in rural Shaanxi in the 1950s," in Barbara Entwisle and Gail Henderson (eds), *Re-Drawing Boundaries: Work, Households, and Gender in China*, Berkeley, CA: University of California Press, 79–96.

Honig, Emily and Hershatter, Gail (1988) "Marriage," in Emily Honig and Gail Hershatter (eds), *Personal Voices: Chinese Women in the 1980s*, Stanford, California: Stanford University Press, 137–166.

Houstourn, M. F., Kramer, R. G., and Barrett, J. M. (1984) "Female predominance of immigration to the United States since 1930: A first look," *International Migration Review*, 18: 908–963.

Hsu, Mei Ling (1994) "The expansion of the Chinese urban system, 1953–1990," *Urban Geography*, 15(6): 514–536.

Hu, Dapeng (2002) "Trade, rural–urban migration, and regional income disparity in developing countries: A spatial general equilibrium model inspired by the case of China," *Regional Science and Urban Economics*, 32(3): 311–338.

Huadong shifan daxue renkou yanjiusuo (East China Normal University Population Research Institute) (2005) "*90 niandai zhongguo renkou fenbu he qianyi liudong di xinxingshi yanjiu*" (Research on population distribution and new trends in migration in China during the 1990s), in National Bureau of Statistics (ed.), *2000 Pucha Guojiaji Zhongdian Keti Yanjiu Baogao* (*National Level Key Research Reports on the 2000 Census*). Beijing: China Statistics Press, 1146–1233 (in Chinese).

Huang, Runlong (2002) "*Jiangsu sheng waila hunjianu di hunyin zhuangtai yu guannian*" (The marriage situation and views of female marriage migrants in Jiangsu), *Renkou Yu Jingji* (*Population and Economics*), 2002(2): 16–21 (in Chinese).

Huang, Xiyi (1999) "Divided gender, divided women: State policy and the labour market," in Jackie West, Minghua Zhao, Xiangqun Chang and Yuan Cheng (eds), *Women of China: Economic and Social Transformation*, London: Macmillan Press, 90–107.

Huang, Youqin (2001) "Gender, hukou, and the occupational attainment of female migrants in China (1985–1990)," *Environment and Planning A*, 33(2): 257–279.

Hugo, Graeme (2003a) "Circular migration: Keeping development rolling?" Migration Policy Institute, [www.migrationinformation.org/Feature/print.cfm?ID=129], June 1 (accessed December 11, 2006).

—— (2003b) "Urbanisation in Asia: An overview," Paper prepared for Conference on African Migration in Comparative Perspective, Johannesburg, South Africa (June 4–7, 2003).

—— (2005) "Migrants in society: Diversity and cohesion," Geneva, Global Commission on International Migration, [www.gcim.org].

—— (2006) "Temporary migration and the labour market in Australia," *Australian Geographer*, 37(2): 211–231.

Humbeck, Eva (1996) "The politics of cultural identity: Thai women in Germany," in Maria Dolors García-Ramon and Janice Monk (eds), *Women of the European Union: The Politics of Work and Daily Life*, London: Routledge, 186–201.

International Organization for Migration (2005) *World Migration 2005: Costs and Benefits of International Migration*, Geneva: International Organization for Migration.

Itzigsohn, Jose (1995) "Migrant remittances, labor markets, and household strategies: A comparative analysis of low-income household strategies in the Caribbean Basin," *Social Forces*, 74(2): 633–655.

Jacka, Tamara (1997) *Women's Work in Rural China: Change and Continuity in an Era of Reform*, Cambridge: Cambridge University Press.

—— (2006) *Rural Women in Contemporary China: Gender, Migration, and Social Change*, Armonk, NY: M. E. Sharpe.

Jacka, Tamara and Song, Xianlin (2003) "My life as a migrant worker (translation)," in Arianne M. Gaetano and Tamara Jacka (eds), *On the Move: Women in Rural-to-Urban Migration in Contemporary China*, Columbia University Press.

Jacka, Tamara and Gaetano, Arianne M. (2004) "Introduction: Focusing on migrant women," in Arianne M. Gaetano and Tamara Jacka (eds), *On the Move: Women in Rural-to-Urban Migration in Contemporary China*, Columbia University Press, 1–38.

Jarvis, Helen (1999) "The tangled webs we weave: Household strategies to co-ordinate home and work," *Work, Employment and Society*, 13(2): 225–247.

Jessop, B. (1999) "Narrating the future of the national economy and the national state? Remarks on remapping regulation and reinventing governance," in G. Steinmetz (ed.), *State/Culture: State Formation After the Cultural Turn*, Ithaca: Cornell University Press, 378–405.

Ji, Ping, Zhang, Kaiti, and Liu, Dawei (1985) "*Beijing jiaoqu nongcun renkou hunyin qianyi qianxi*" (Marriage migration in Beijing's rural suburbs), *Zhongguo Shehui Kexue* (*Social Sciences in China*), 1985(3): 201–213 (in Chinese).

Jian, Xinhua and Zhang, Jianwei (2005) "*Cong 'mingong chao' dao 'mingong huang': nongcun shengyu laodongli youxiaozhuanyi de zhidu fenxi*" (From "the wave of migrants" to "the shortage of migrants": the institutional analysis of the effective transfer of rural surplus labor), *Renkou Yanjiu* (*Population Research*), 29(2): 49–55 (in Chinese).

Jiao, Jianquan (2002) "A study on floating population from rights point of view," *Renkou Yu Jingji* (*Population and Economics*), 2002(3): 73–75.

Johnson, D. Gale (2002) "Can agricultural labour adjustment occur primarily through creation of rural non-farm jobs in China?" *Urban Studies*, 39(12): 2163–2174.

Johnson, Kay Ann (1983) *Women, the Family, and Peasant Revolution in China*, Chicago: University of Chicago Press.

Jolivet, Muriel (1997) *Japan: The Childless Society*, London: Routledge.

Katz, Eliakim and Stark, Oded (1986) "Labor migration and risk aversion in less developed countries," *Journal of Labor Economics*, 4(1): 134–149.

King, Russell (1986) "Return migration and regional economic problems," in Russell King (ed.), *Return Migration and Regional Economic Development: An Overview*, Dover, New Hampshire: Croom Helm.

Knight, John and Song, Lina (1995) "Towards a labour market in China," *Oxford Review of Economic Policy*, 11(4): 97–117.

—— (1999) *The Rural–Urban Divide: Economic Disparities and Interactions in China*, Oxford: Oxford University Press.

Knight, John, Song, Lina and Huaibin, J. (1999) "Chinese rural migrants in urban enterprises: Three perspectives," *Journal of Development Studies*, 35(3): 73–104.

Kornai, Janos (1997) *Struggle and Hope: Essays on Stabilization and Reform in a Post-Socialist Economy*, Cheltenham, UK: Edward Elgar.

Kwong, Julia (2004) "Educating migrant children: Negotiations between the state and civil society," *China Quarterly*, 180: 1073–1088.

Lavely, William (1991) "Marriage and mobility under rural collectivization," in Rubie S. Watson and Patrick Buckley Ebrey (eds), *Marriage and Inequality in Chinese Society*, Berkeley: University of California Press, 286–312.

Lavely, William and Mason, William M. (2006) "An evaluation of the one percent clustered sample of the 1990 census of China," *Demographic Research*, 15: 329–346.

Lawson, Victoria A. (1998) "Hierarchical households and gendered migration in Latin America: Feminist extensions to migration research," *Progress in Human Geography*, 22(1): 39–53.

Lee, Ching Kwan (1995) "Engendering the worlds of labor: Women workers, labor markets, and production politics in the South China economic miracle," *American Sociological Review*, 60: 378–397.

—— (1998) *Gender and the South China Miracle: Two Worlds of Factory Women*, Berkeley, CA: University of California Press.

—— (2007) *Against the Law: Labor Protests in China's Rustbelt and Sunbelt*, Berkeley: University of California Press.

Lee, On-Jook (1984) *Urban-to-Rural Return Migration in Korea*, Seoul: Seoul National University Press.

Lewis, W. A. (1954) "Economic development with unlimited supplies of labour," *Manchester School of Economic and Social Studies*, May 1954: 131–191.

Li, Jiang Hong and Lavely, William (1995) "Rural economy and male marriage in China: Jurong, Jiangsu 1933," *Journal of Family History*, 20(3): 289–306.

Li, Ling (1994a) *Gender and Development: Innercity and Suburban Women in China*, Guangzhou, China: Centre for Urban and Regional Studies, Zhongshan University, Working Paper No. 11.

—— (1994b) *Migration, Urbanization and Urban Planning in Guangdong Province*, Guangzhou, China: Center for Urban and Regional Studies, Zhongshan University, Working Paper No. 8.

Li, Shi (1999) "Effects of labor out-migration on income growth and inequality in rural China," *Development and Society*, 28(1): 93–114.

—— (2003) "Zhongguo geren shouru fenpei yanjiu huigu yu zhanwang" (A review of income inequality in China), *Jijixue Jikan (China Economic Quarterly)*, 2(2): 379–404 (in Chinese).

Li, Shuzhuo (1993) "*Bashi niandai zhongguo renkou qianxi de xingbie chayi yanjiu*" (A gender-difference research on population migration of China in 1980s), *Renkou Xuekan (Population Journal)*, 1993(5): 14–19 (in Chinese).

Li, Si-Ming (1995) "Population mobility and urban and rural development in Mainland China," *Issues and Studies*, 31(9): 37–54.

—— (1997) "Population migration, regional economic growth and income determination: A comparative study of Dongguan and Meizhou, China," *Urban Studies*, 34(7): 999–1026.

—— (2004) "Population migration and urbanization in China: A comparative analysis of the 1990 population census and the 1995 national one percent sample population survey," *International Migration Review*, 38(2): 655–685.

Li, Si-Ming and Siu, Yat-Ming (1994) "Population mobility," in Y. M. Yeung and David K. Y. Chu (eds), *Guangdong: Survey of a Province Undergoing Rapid Changes,* Hong Kong: Chinese University Press, 373–400.

—— (2002) "Fertility of migrants and non-migrants in Dongguan and Meizhou: A study of the impact of regional development and inter-regional migration on fertility behaviour in China," in Kwan-yiu Wong and Jianfa Shen (eds), *Resource Management, Urbanization and Governance in Hong Kong and the Zhujiang Delta*, Hong Kong: Chinese University Press, 34–53.

Li, Xiao (2004) "Finding balance: Meeting the needs of migrant kids," [www.chinagate.com.en/english/10249.htm], June 23 (accessed April 12, 2007).

Lian, Xiaomei (2002) "Tentative analyses on the influence of population floating on the coordinative economic development between regions," *Renkou Xuekan (Population Journal)*, 2002(4): 41–45 (in Chinese).

Liang, Zai (1999) "Foreign investment, economic growth, and temporary migration: The

case of Shenzhen special economic zone, China," *Development and Society*, 28(1): 115–137.
—— (2001a) "The age of migration in China," *Population and Development Review*, 27(3): 499–528.
—— (2001b) "Demography of illicit emigration from China: A sending country's perspective," *Sociological Forum*, 16(4): 677–701.
Liang, Zai and White, Michael J. (1997) "Market transition, government policies, and interprovincial migration in China: 1983–1988," *Economic Development and Cultural Change*, 45(2): 19–37.
Liang, Zai and Wu, Yingfeng (2003) "Return migration in China: New methods and findings," Paper presented at the 2003 Annual Meeting of the Population Association of America, Minneapolis, Minnesota.
Liang, Zai and Ma, Zhongdong (2004) "China's floating population: New evidence from the 2000 census," *Population and Development Review*, 33(3): 467–488.
Liang, Zai and Chen, Yiu Por (2007) "The educational consequences of migration for children in China," *Social Science Research*, 36(1): 285–247.
Liaowan xinwen zhoukan (*Outlook*) (2007) "*Huji gaige buneng yiqian liaozhi ying zhuoyan chengshi zhonghe peitao gaige*" (Hukou reform is more than moving and should focus on comprehensive adjustment reform), [http://unn.people.com.cn/GB/14748-5259895.html], January 9 (accessed May 8, 2007) (in Chinese).
Lievens, John (1999) "Family-forming migration from Turkey and Morocco to Belgium: The demand for marriage partners from the countries of origin," *International Migration Review*, 33(3): 717–744.
Lin, George C. S. (1997) *Red Capitalism in South China: Growth and Development of the Pearl River Delta*, Vancouver: University of British Columbia Press.
Lin, George C. S. and Ho, Samuel P. S. (2005) "The state, land system, and land development processes in contemporary China," *Annals of the Association of American Geographers*, 95(2): 411–436.
Lin, Justin Y. (1992) "Rural reforms and agricultural growth in China," *American Economic Review*, 82: 34–51.
Lin, Nan (2000) "Understanding the social inequality system and family and household dynamics in China," in Barbara Entwisle and Gail E. Henderson (eds), *Re-Drawing Boundaries: Work, Households, and Gender in China*, Berkeley, CA: University of California Press, 295–303.
Lingam, Lakshmi (2005) "Structural adjustment, gender and household survival strategies: Review of evidences and concerns," Center for the Education of Women, the University of Michigan [www.cew.umich.edu].
Liu, Alan P. L. (1991) "Economic reform, mobility strategies, and national integration in China," *Asian Survey*, 31(5): 393–408.
Liu, Ta and Chan, Kam Wing (2001) "National statistics on internal migration in China: Comparability problems," *China Information*, 15(2): 75–113.
Liu, Xiankang (1990) "*Guanyui Xiaoshanshi 'wailainu' zhuangkuang ji qi guan li wenti*" (The conditions and management problems of "women from outside" in Xiaoshan city), *Renkou Yanjiu* (*Population Research*), 14(6): 31–36 (in Chinese).
Lo, Chor Pang (1994) "Economic reforms and socialist city structure: A case study of Guangzhou, China," *Urban Geography*, 15: 128–149.
Logan, John R. (2002) "Three challenges for the Chinese city: Globalization, migration, and market reform," in John R. Logan (ed.), *The New Chinese City: Globalization and Market Reform*, Malden, MA: Blackwell, 3–21.

Los Angeles Times (2004) "China's migrant workers ask for little and receive nothing," January 21, p. A4.

Lou, Binbin, Zheng, Zhenzhen, Connelly, Rachel and Roberts, Kenneth (2004) "The migration experiences of young women from poor counties in Sichuan and Anhui," in Arianne M. Gaetano and Tamara Jacka (eds), *On the Move: Women in Rural-to-Urban Migration in Contemporary China*, Columbia University Press, 207–242.

Loughlin, Philip H. and Pannell, Clifton W. (2001) "Growing economic links and regional development in the Central Asian Republics and Xinjiang, China," *Post-Soviet Geography and Economics*, 42(7): 469–90.

Lu, Li (1997) "*Funu jingji diwei yu funu renli ziben guanxi de shizhen yanjiu*" (Women's economic status and their human capital in China), *Renkou Yanjiu* (*Population Research*), 21(2): 50–54 (in Chinese).

Lu, Xianghu and Wang, Yonggang (2006) "*Zhongguo 'xiang-cheng' renkou qianyi guimo de cesuan yu fenxi* (1979–2003)" (Estimation and analysis on Chinese rural–urban migration size), *Xibei Renkou* (*Northwest Population*), 2006(1): 14–16 (in Chinese).

Lucas, R. E. B. and Stark, O. (1985) "Motivations to remit: Evidence from Botswana," *Journal of Political Economy*, 93(5): 901–918.

Ma, Lanmei, Chen, Zhongmin and Du, Guizhen (1995) "*Dui 'wailaimei' hunyu guanli qingkuang de diaocha yu sikao*" (Investigation and contemplation of the fertility management of female inmigrants), *Renkou Yanjiu* (*Population Research*), 10(1): 56–58 (in Chinese).

Ma, Laurence J. C. (1996) "The spatial patterns of interprovincial rural-to-urban migration in China, 1982–1987," *Chinese Environment and Development*, 7: 73–102.

Ma, Laurence J. C. and Cui, Gonghao (1987) "Administrative changes and urban population in China," *Annals of the Association of American Geographers*, 77(3): 373–395.

Ma, Laurence J. C. and Xiang, Biao (1998) "Native place, migration and the emergence of peasant enclaves in Beijing," *China Quarterly*, 155: 546–581.

Ma, Zhongdong (2001) "Urban labour-force experience as a determinant of rural occupation change: Evidence from recent urban–rural return migration in China," *Environment and Planning A*, 33(2): 237–255.

—— (2002) "Social capital mobilization and income returns to entrepreneurship: The case of return migration in rural China," *Environment and Planning A*, 34(10): 1763–1784.

Ma, Zhongdong, Liaw, K.-L., and Zeng, Y. (1997) "Migrations in the urban–rural hierarchy of China: Insights from the microdata of the 1987 national survey," *Environment and Planning A*, 29(4): 707–730.

Ma, Zhongdong, Liang, Zai, Zhang, Weimin, and Cui, Hongyan (2004) "*Laodongli liudong: zhongguo nongcun shouru zengzhang di xinyinsu*" (Labor migration as a new determinant of income growth in rural China), *Renkou Yanjiu* (*Population Research*), 28(3): 2–10 (in Chinese).

McGee, Terence G. (1982) "Labor markets, urban systems, and the urbanization process in Southeast Asian countries," *Papers of the East-West Population Institute*, 81: 1–28.

Mallee, H. (1996) "In defense of migration: Recent Chinese studies on rural population mobility," *China Information*, 10(3/4): 108–140.

Mann, Susan (2000) "Work and household in Chinese culture: Historical perspectives," in Barbara Entwisle and Gail Henderson (eds), *Re-Drawing Boundaries: Work, Households, and Gender in China*, Berkeley, CA: University of California Press, 15–32.

Massey, Douglas S. (1990) "The social and economic origins of immigration," *Annals of the American Academy of Political and Social Science*, 510: 60–72.

Massey, Douglas S. and Espinosa, Kristin E. (1997) "What's driving Mexico–US migra-

tion? A theoretical, empirical, and policy analysis," *American Journal of Sociology*, 102(4): 939–999.
Massey, Douglas S., Arango, Joaquin, Hugo, Graeme, Kouaouci, Ali, Pellegrino, Adela, and Taylor, J. Edward (1993) "Theories of international migration: A review and appraisal," *Population and Development Review*, 19: 431–466.
Maurer-Fazio, Margaret (1995) "Building a labor market in China," *Current History*, 94(593): 285–289.
Maurer-Fazio, Margaret, Rawski, Thomas G., and Zhang, Wei (1999) "Inequality in the rewards for holding up half the sky: Gender wage gaps in China's urban labor markets, 1988–1994," *China Journal*, 41: 55–88.
Meng, Jianjun (2003) "Population migration and economic development in China," [www1.msh-paris.fr/reseauemploi/Shanghai/JianJunMengShanghai/JianJunShanghai1.html] (accessed July 7, 2004).
Meng, Xin (2000) *Labor Market Reform in China*, Cambridge: Cambridge University Press.
Messner, Steven F., Liu, Jianhong, and Karstedt, Susanne (2007) "Economic reform and crime in contemporary China: Paradoxes of a planned transition," in John Logan (ed.), *Urban China in Transition*, Blackwell, forthcoming.
Michelson, Ethan and Parish, William L. (2000) "Gender differences in economic success: Rural China in 1991," in Barbara Entwisle and Gail Henderson (eds), *Re-Drawing Boundaries: Work, Households, and Gender in China*, Berkeley, CA: University of California Press, 134–156.
Min, Han and Eades, J. S. (1995) "Brides, bachelors and brokers: The marriage market in rural Anhui in an era of economic reform," *Modern Asian Studies*, 29(4): 841–869.
Mincer, Jacob (1978) "Family migration decisions," *Journal of Political Economy*, 86(5): 749–773.
Mitchneck, Beth and Plane, David A. (1995a) "Migration patterns during a period of political and economic shocks in the Former Soviet Union: A case study of Yaroslavy' Oblast," *Professional Geographer*, 47(1): 17–30.
—— (1995b) "Migration and the quasi-labor market in Russia," *International Regional Science Review*, 18(3): 267–288.
Momsen, Janet (1992) "Gender selectivity in Caribbean migration," in S. Chant (ed.), *Gender and Migration in Developing Countries*, London: Belhaven Press, 73–90.
Montlake, Simon (2006) "China's factories hit an unlikely shortage: Labor," *Christian Science Monitor*, [www.csmonitor.com/2006-0501/p01s03-woap.html], May 1 (accessed May 7, 2007).
Mulder, Clara H. and Wagner, Michael (1993) "Migration and marriage in the life course: A method for studying synchronized events," *European Journal of Population*, 9(1): 55–76.
Murphy, Rachel (1999) "Return migrants and economic diversification in two counties in South Jiangxi, China," *Journal of International Development*, 11: 661–672.
—— (2000) "Return migration, entrepreneurship and local state corporatism in rural China: The experience of two counties in south Jiangxi," *Journal of Contemporary China*, 9(24): 231–247.
—— (2002) *How Migrant Labor Is Changing Rural China*, Cambridge: Cambridge University Press.
Nagar, Richa, Lawson, Vicky, McDowell, Linda, and Hanson, Susan (2002) "Locating globalization: Feminist (re)readings of the subjects and spaces of globalization," *Economic Geography*, 78(3): 257–284.

NBS (National Bureau of Statistics) (1999) *Xinzhongguo Wushinian Tongji Ziliao Huibian* (*Comprehensive Statistical Data and Materials on 50 Years of New China*), Beijing: Zhongguo tongji chubanshe (China Statistics Press) (in Chinese).
—— (2002) *Zhongguo 2000 Nian Renkou Pucha Ziliao* (*Tabulation on the 2000 Population Census of the People's Republic of China*), Vols I and III, Beijing: Zhongguo tongji chubanshe (China Statistics Press) (in Chinese).
—— (2006) "*2005 nian quanguo 1% renkou chouyang diaocha zhuyao shuju gongbao*" (Announcement of major statistics from the 2005 National One Percent Population Sample Survey), [www.cpirc.org.cn/tjsj/tjsj_cy_detail.asp?id=6628] (in Chinese).
Nelson, Joan M. (1976) "Sojourners versus urbanites: Causes and consequences of temporary versus permanent cityward migration in developing countries," *Economic Development and Cultural Change*, 24(4): 721–757.
Newbold, K. Bruce (2001) "Counting migrants and migrations: Comparing lifetime and fixed-interval return and onward migration," *Economic Geography*, 77(1): 23–40.
Newsweek (2005) "China's century (special report)," May 9.
Nongyebu nongcun jingji yanjiu zhongxin (NNJYZ) (Ministry of Agriculture) (1995) *Zhongguo Nongcun Laodongli Liudong: Gean Fangtan Ziliao* (*Labor Migration in China: Accounts of Individual Interviews*), Ministry of Agriculture (in Chinese).
—— (1999) *Zhongguo Nongcun Laodongli Huilui Yanjiu: Gean Fangtan Ziliao* (*Return Labor Migration in China: Accounts of Individual Interviews*), Ministry of Agriculture (in Chinese).
Oberai, A. S. and Singh, Manmohan (1983) *Causes and Consequences of Internal Migration: A Study of the Indian Punjab*, New Delhi: Oxford University Press.
O'Brien, Kevin J. and Li, Lianjiang (2006) *Rightful Resistance in Rural China*, Cambridge: Cambridge University Press.
Odland, John and Ellis, Mark (1988) "Household organization and the interregional variation of out-migration rates," *Demography*, 25(4): 567–579.
Ong, Aihwa (1999) *Flexible Citizenship: The Cultural Logics of Transnationality*, Durham: Duke University Press.
Ortiz, Vilma (1996) "Migration and marriage among Puerto Rican women," *International Migration Review*, 30(2): 460–484.
Pannell, Clifton W. and Ma, Laurence J. C. (1997) "Urban transition and interstate relations in a dynamic post-Soviet borderland: The Xingjiang Uygur Autonomous Region of China," *Post-Soviet Geography and Economics*, 38(4): 206–229.
Park, Kyung Ae (1992) *Women and Revolution in China: The Sources of Constraints on Women's Emancipation*, Michigan University, Franklin and Marshall College, Working Paper No. 230.
Peck, Jamie (1994) "Regulating labour: The social regulation and reproduction of local labour-markets," in Ashe Amin and Nigel Thrift (eds), *Globalization, Institutions and Regional Development in Europe*, Oxford: Oxford University Press, 147–176.
Peng, Xizhe, Zheng, Guizhen and Chen, Yuexin (1998) "Wailai nuxing laodongli renkou chuxian di xinqingkuang" (New circumstances of female migrant labor force: A survey of two districts in Shanghai), *Nanfang Renkou* (*Southern Population*), 1998(2): 37–40 (in Chinese).
Pilkington, Hilary (1998) *Migration, Displacement and Identity in Post-Soviet Russia*, New York: Routledge.
Piore, M. (1979) *Birds of Passage: Migrant Labour in Industrial Societies*, New York: Cambridge University Press.

Piper, Nicola (1999) "Labor migration, trafficking and international marriage: Female cross-border movements into Japan," *Asian Journal of Women's Studies*, 5(2): 69–99.
Poncet, Sandra (2006) "Provincial migration dynamics in China: Borders, costs and economic motivations," *Regional Science and Urban Economics*, 36(3): 385–398.
Potter, S. and Potter, J. (1990) *China's Peasants: The Anthropology of a Revolution*, Cambridge: Cambridge University Press.
Pun, Ngai (2005) *Made in China: Women Factory Workers in a Global Workplace*, Durham, NC: Duke University Press.
Qiu, Haiying (2001) "Nongcun laodongli huiliu yu laodongli miji xing chanye di kaifa" (Return labor migration and the development of labor-intensive industry), *Renkou Xuekan (Population Journal)*, 2001(3): 52–55 (in Chinese).
Qiu, Ziyi, Xie, Ping, and Zhou, Fanglian (2004) "Renkou liudong dui jingji shehui fazhan di yingxiang" (Impacts of migration on socioeconomic development), *Renkou Xuekan (Population Journal)*, 2004(1): 47–52 (in Chinese).
Radcliffe, Sarah A. (1990) "Between hearth and labour market: The recruitment of peasant women in the Andes," *International Migration Review*, 24(2): 229–249.
—— (1991) "The role of gender in peasant migration: Conceptual issues from the Peruvian Andes," *Review of Radical Political Economics*, 23(3 and 4): 129–147.
Ravenstein, E. G. (1885) "The laws of migration," *Journal of the Statistical Society*, 48: 167–227.
Renmin ribao (People's Daily) (1989) "Duoshu nongmin tonghunchuan bu chaoguo 25 gongli" (Most rural people's marriage boundaries are less than 25 km), August 11: 4 (in Chinese).
Reyes, Belinda I. (1997) *Dynamics of Immigration: Return Migration to Western Mexico*, CA: Public Policy Institute of California.
Riley, Nancy E. (1996) "China's 'missing girls': Prospects and policy," *Population Today*, 24(2): 4–5.
Riley, Nancy E. and Gardner, Robert W. (1993) "Migration decisions: The role of gender," in United Nations Department for Economic and Social Information and Policy Analysis (ed.), *Internal Migration of Women in Developing Countries*, Proceedings of the United Nations Expert Meeting on the Feminization of Internal Migration, Aguascalientes, Mexico, October 22–25 1991, New York: United Nations, 195–206.
Roberts, Kenneth D. (1997) "China's 'tidal wave' of migrant labor: What can we learn from Mexican undocumented migration to the United States?" *International Migration Review*, 31(2): 249–293.
—— (2007) "The changing profile of Chinese labor migration," in Zhongwei Zhao and Fei Guo (eds), *Transition and Challenge: China's Population at the Beginning of the 21st Century*, Oxford University Press, 233–250.
Rosenzweig, Mark R. and Stark, Oded (1989) "Consumption smoothing, migration, and marriage: Evidence from rural India," *Journal of Political Economy*, 97(4): 905–926.
Rowland, Donald T. (1994) "Family characteristics of the migrants," in Lincoln H. Day and Xia Ma (eds), *Migration and Urbanization in China*, Armonk, New York: M. E. Sharpe, 129–154.
Rozelle, Scott, Guo, Li, Shen, Minggao, Hughart, Amelia, and Giles, John (1999) "Leaving China's farms: Survey results of new paths and remaining hurdles to rural migration," *China Quarterly*, 158: 367–393.
Saxenian, AnnaLee (2005) "From brain drain to brain circulation: Transnational communities and regional upgrading in India and China," *Studies in Comparative International Development*, 40(2): 35–61.

Scharping, Thomas (ed.) (1997) *Floating Population and Migration in China: The Impact of Economic Reforms*, Hamburg: Institut für Asienkunde.

Schultz, T. W. (1961) "Investment in human capital," *American Economic Review*, 51: 1–17.

Shen, Jianfa (1999) "Modelling regional migration in China: Estimation and decomposition," *Environment and Planning A*, 31(7): 1223–1238.

—— (2002) "A study of the temporary population in Chinese cities," *Habitat International*, 26(3): 363–377.

Shen, Xiaoping (2001) "Regional variations and changes in industrial productivity in China, 1980–1995," *Asian Geographer*, 20(1 and 2): 53–78.

Shen, Yimin and Tong, Chengzhu (1992) *Zhongguo Renkou Qianyi (China's Population Migration)*, Beijing: Zhongguo tongji chubanshe (in Chinese).

Sichuan Statistical Bureau (1996) *Sichuan Tongji Nianjian 1996 (Sichuan Statistical Yearbook 1996)*, Beijing: Zhongguo tongji chubanshe (in Chinese).

Sicular, T., Yue, X., Gustafsson, B., and Li, S (2007) "The urban–rural income gap and inequality in China," *Review of Income and Wealth*, 53(1): 93–126.

Silvey, Rachel and Lawson, Victoria (1999) "Placing the migrant," *Annals of the Association of American Geographers*, 89(1): 121–132.

Simons, A. B. and Cardona, G. R. (1972) "Rural–urban migration: Who comes, who stays, who returns? The case of Bogota, Colombia, 1928–68," *International Migration Review*, 6: 166–181.

Sjaastad, L. A. (1962) "The costs and returns of human migration," *Journal of Political Economy*, 70 (Supplement): 80–93.

Skinner, G. William (1976) "Mobility strategies in late imperial China: A regional systems analysis," in Carol A. Smith (ed.), *Regional Analysis, Vol I: Economic Systems*, New York: Academic Press, 327–364.

Skocpol, Theda (1985) "Bringing the state back in: Strategies of analysis in current research," in Peter B. Evans, Dietrich Rueschemeyer and Theda Skocpol (eds), *Bringing the State Back In*, Cambridge: Cambridge University Press, 1–37.

Smart, Alan and Smart, Josephine (2001) "Local citizenship: Welfare reform urban/rural status, and exclusion in China," *Environment and Planning A*, 33(10): 1853–1869.

Smith, Christopher J. (1996) "Migration as an agent of change in contemporary China," *Chinese Environment and Development*, 7(1 and 2): 14–55.

—— (2000) *China in the Post-Utopian Age*, Boulder, CO: Westview Press.

Solinger, Dorothy J. (1995) "The floating population in the cities: Chances for assimilation?" in Deborah Davis, Richard Kraus, Barry Naughton and Elizabeth J. Perry (eds), *Urban Spaces in Contemporary China*, Cambridge: Cambridge University Press, 113–139.

—— (1999a) *Contesting Citizenship in Urban China: Peasant Migrants, the State, and the Logic of the Market*, Berkeley: University of California Press.

—— (1999b) "Citizenship issues in China's internal migration: Comparisons with Germany and Japan," *Political Science Quarterly*, 114(3): 455–478.

Song, Guochen and Gu, Chaolin (1999) "Beijing nuxing liudong renkou di jiating leixing ji qi xingcheng yinsu" (The factors and household types of female migrants in Beijing), *Renwen Dili (Human Geography)*, 1999(2): 11–14 (in Chinese).

Speare, A. Jr and Goldscheider, F. K. (1987) "Effects of marital status change on residential mobility," *Journal of Marriage and the Family*, 49: 455–464.

SSB (State Statistical Bureau) (1992) *Zhongguo Renkou Tongji Nianjian (China Population Statistical Yearbook) 1992*, Beijing: Zhongguo tongji chubanshe (China Statistical Publishing House) (in Chinese).

—— (1993) *Zhongguo 1990 Nien Renkou Pucha Ziliao* (*Tabulation on the 1990 Population Census of the People's Republic of China*), Vol IV, Beijing: Zhongguo tongji chubanshe (in Chinese).
Stark, David (1986) "Rethinking internal labor markets: New insights from a comparative perspective," *American Sociological Review*, 51: 492–504.
Stark, Oded and Lucas, Robert E. B. (1988) "Migration, remittances, and the family," *Economic Development and Cultural Change*, 36(3): 465–481.
Stinner, William F., Xu, Wu, and Wei, Jin (1993) "Migrant status and labor market outcomes in urban and rural Hebei Province, China," *Rural Sociology*, 58(3): 366–386.
Stockman, Norman (1994) "Gender inequality and social structure in urban China," *Sociology*, 28(3): 759–777.
Sun, Wenlan (1991) *Lihun Zai Zhongguo* (*Divorce in China*), Beijing: Zhongguo funu chubanshe (in Chinese).
Szelenyi, Ivan and Kostello, Eric (1996) "The market transition debate: Toward a synthesis?," *American Journal of Sociology*, 101(4): 1082–1096.
Tam, Siumi Maria (2000) "Modernization from a grassroots perspective: Women workers in Shekou Industrial Zone," in Si-ming Li and Wing-shing Tang (eds), *China's Regions, Polity, and Economy*, Hong Kong: Chinese University Press, 371–390.
Tan, Shen (1996a) "*Zhongguo nongcun laodongli liaodong di xingbie cha*" (Gender differences in the migration of rural labor force), Paper presented at the International Conference on Rural Labor Migration, Beijing, China (in Chinese).
—— (1996b) "The process and achievements of the study on marriage and family in China," *Marriage and Family Review*, 22(1–2): 19–53 (in Chinese).
Tang, Wing Shing (1991) *Regional Uneven Development in China, with Special Reference to the Period Between 1978 and 1988*, Hong Kong: Chinese University of Hong Kong, Department of Geography occasional paper no. 110.
Tang, Wing Shing, Chu, David K. Y., and Fan, C. Cindy (1993) "Economic reform and regional development in China in the 21st century," in Yue Man Yeung (ed.), *Pacific Asia in the 21st Century: Geographical and Developmental Perspectives*, Hong Kong: Chinese University Press, 105–133.
Tang, Xuemei (1993) "*Beijing shi renkou qianyi he renkou liudong*" (Population migration and mobility in Beijing), *Renkou Yanjiu* (*Population Research*), 17(4): 52–55 (in Chinese).
Taubmann, Wolfgang (1997) "Migration into rural towns (zhen) – Some results of a research project on rurban urbanisation in China," in Thomas Scharping (ed.), *Floating Population and Migration in China: The Impact of Economic Reforms*, Hamburg: Mitteilungen Des Instituts Für Asienkunde, 236–263.
Thadani, Veena N. and Todaro, Michael P. (1984) "Female migration: A conceptual framework," in James T. Fawcett, Siew-Ean Khoo, and Peter C. Smith (eds), *Women in the Cities of Asia: Migration and Urban Adaptation*, Boulder, Colorado: Westview Press.
Todaro, Michael P. (1969) "A model of labor migration and urban unemployment in less developed countries," *American Economic Review*, 59: 138–148.
—— (1976) *Internal Migration in Developing Countries: A Review of Theory, Evidence, Methodology, and Research Priorities*, Geneva: International Labour Office.
Tyson, Ann and Tyson, James (1996) "China's human avalanche," *Current History*, 95(602): 277–283.
United Nations (2007) *World Population Prospects: The 2006 Revision*, [www.un.org/esa/population/publications/wpp/2006/wpp/2006.htm] (accessed May 9, 2007).

References

United Nations Secretariat (1993) "Types of female migration," in Department for Economic and Social Information and Policy Analysis, United Nations (ed.), *Internal Migration of Women in Developing Countries,* Proceedings of the United Nations Expert Meeting on the Feminization of Internal Migration, Aguascalientes, Mexico, October 22–25 1991, New York: United Nations, 94–115.

Veeck, Gregory, Pannell, Clifton W., Smith, Christopher J., and Huang, Youqin (2007) *China's Geography: Globalization and the Dynamics of Political, Economic, and Social Change,* Lanham, Maryland: Rowman & Littlefield.

Walby, Sylvia (1990) *Theorizing Patriarchy,* Oxford: Blackwell.

Waldinger, Roger (1992) "Taking care of the guests: The impact of immigrants on services," *International Journal of Urban and Regional Research,* 16(1): 97–113.

Wallace, Clare (2002) "Household strategies: Their conceptual relevance and analytical scope in social research," *Sociology,* 36(2): 275–292.

Wan, Chuan (2001) "A commentary on Professor Zhang Qing-Wu's academic viewpoints of household register administration," *Renkou Yu Jingji (Population and Economics),* 2001(4): 75–79 (in Chinese).

Wang, Beiyu and Li, Lulu (2001) *Zhongguo Chengshi Laodongli Liudong (Migration of Urban Labor in China),* Beijing: Beijing chubanshe (in Chinese).

Wang, De and Ye, Hui (2004) "*1990 nian yihou di zhongguo renkou qianyi yanjiu zhongshu*" (A review of research on migration in China since 1990), *Renkou Xuekan (Population Journal),* 2004(1): 40–46 (in Chinese).

Wang, Fang (2005) "*Zhongguo chengzhenhua jingcheng zhong de liudong renkou zinu shou jiaoyu wenti*" (The issues on the education of migrant children during the process of urbanization in China), *Fazhan Zhanlue (Development Strategy),* 2005(9): 27–31 (in Chinese).

Wang, Fei-Ling (2005) *Organizing Through Division and Exclusion: China's Hukou System,* Stanford, CA: Stanford University Press.

Wang, Feng (1997) "The breakdown of a great wall: Recent changes in the household registration system of China," in Thomas Scharping (ed.), *Floating Population and Migration in China: The Impacts of Economic Reforms,* Hamburg: Institute of Asian Studies, 149–165.

—— (2000) "Gendered migration and the migration of genders in contemporary China," in Barbara Entwisle and Gail Henderson (eds), *Re-Drawing Boundaries: Work, Households, and Gender in China,* Berkeley: University of California Press, 231–242.

Wang, Feng, Zho, Xuejin and Ruan, Danching (2002) "Rural migrants in Shanghai: Living under the shadow of socialism," *International Migration Review,* 36(2): 520–545.

Wang, Guixin (1992) "*Shanghai shengji renkou qianyi yu juli guanxi zhi tantao*" (Relationships between interprovincial migration and the migration distance in Shanghai), *Renkou Yanjiu (Population Research),* 16(4): 1–7 (in Chinese).

—— (2001) "*21 shiji renkou qianyi jiang tuidong zhongguo xiandaihua jiasu fazhan*" (Migration in the 21st century will propel the modernization in China forward more rapidly), *Renkou Xuekan (Population Journal),* 2001(5): 31–33 (in Chinese).

—— (2004) "*Gaige kaifang yilai zhongguo renkou qianyi fazhan di jige tezheng*" (Several characteristics of population migration in China since the economic reforms), *Renkou Yu Jingji (Population and Economics),* 2004(4): 1–8, 14 (in Chinese).

Wang, Guixin, Gao, Hui, Xu, Wei and Chen, Guoxiang (2002) "*Xiao chengzhen wailai laodongli jiben zhuangkuang ji dui xiao chengzhen fazhan yingxiang fenxi*" (Analyses on the situation of the labor force and its influences on the development of the little towns), *Renkou Xuekan (Population Journal),* 2002(3): 3–7 (in Chinese).

Wang, Jianming and Hu, Qi (1996) *Zhongguo Liudong Renkou (Floating Population in China)*, Shanghai: Shanghai caijing daxue chubanshe (Shanghai Finance University Press) (in Chinese).

Wang, Wenfei Winnie and Fan, C. Cindy (2006) "Success or failure: Selectivity and reasons of return migration in Sichuan and Anhui, China," *Environment and Planning A*, 38(5): 939–958.

Wang, Wenlu (2003) "*Renkou chengzhenhua beijingxia di huji zhidu bianqian: Shijiazhuang shi huji zhidu gaige anli fenxi*" (Changes in household registering system against the urbanization background: Shijiazhuang city as a case), *Renkou Yanjiu (Population Research)*, 27(6): 8–13.

Wang, Xiyu, Cui, Chuanyi, and Zhao, Yang (2003) "*Da gong yu hui xiang: jiuye zhuangbian he nong cun fazhan*" (Migrate or return: the change of employment and the development of rural areas), *Management World (Guanli Shijie)*, 2003(7): 99–109 (in Chinese).

Watts, Susan J. (1983) "Marriage migration, a neglected form of long-term mobility: a case study from Ilorin, Nigeria," *International Migration Review*, 17(4): 682–698.

Wei, Yehua Dennis (2000) *Regional Development in China: States, Globalization, and Inequality*, London: Routledge.

West, Loraine A. and Zhao, Yaohui (eds) (2000) *Rural Labor Flows in China*, Berkeley, California: Institute of East Asian Studies.

White, Lynn T. (1977) "Deviance, modernization, rations, and household register in urban China," in Amy Auerbacher Wilson, Sidney Leonard Greenblatt, and Richard Whittingham Wilson (eds), *Deviance and Social Control in Chinese Society,* New York: Praeger, 151–172.

Whyte, Martin King (2000) "The perils of assessing trends in gender inequality in China," in Barbara Entwisle and Gail Henderson (eds), *Re-Drawing Boundaries: Work, Households, and Gender in China*, Berkeley, CA: University of California Press, 157–170.

Willekens, Frans (1987) "Migration and development: A micro-perspective," *Journal of Institute of Economic Research*, 22(2): 51–68.

Willis, Katie and Yeoh, Brenda (2000) "Gender and transnational household strategies: Singaporean migration to China," *Regional Studies*, 34(3): 253–264.

Wolf, Magery (1972) *Women and the Family in Rural Taiwan*, Stanford, California: Stanford University Press.

Wolpert, Julian (1965) "Behavioral aspects of the decision to migrate," *Papers and Proceedings of the Regional Science Association*, 15: 159–169.

Wong, Linda and Huen, Wai-Po (1998) "Reforming the household registration system: A preliminary glimpse of the blue chop household registration system in Shanghai and Shenzhen," *International Migration Review*, 32(4): 974–994.

Woon, Yuen-Fong (1994) "Family strategies of prosperous peasants in an emigrant community in South China: A three-year perspective (1988–1991)," *Canadian Journal of Development Studies,* 15(1): 10–33.

Wright, Richard and Ellis, Mark (1997) "Nativity, ethnicity, and the evolution of the intra-urban division of labor in metropolitan Los Angeles," *Urban Geography*, 18: 243–263.

—— (2000) "The ethnic and gender division of labor compared among immigrants to Los Angeles," *International Journal of Urban and Regional Research*, 24(3): 583–600.

Wu, Fulong and Ma, Laurence J. C. (2005) "The Chinese city in transition: Towards theorizing China's urban restructuring," in Laurence J. C. Ma and Fulong Wu (eds),

Restructuring the Chinese City: Changing Society, Economy and Space, London: Routledge, 260–279.

Wu, Harry Xiaoying (1994) "Rural to urban migration in the People's Republic of China," *China Quarterly*, 139: 669–698.

Wu, Jieh-min (2000) "Launching satellites: Predatory land policy and forged industrialization in interior China," in Si-ming Li and Wing-shing Tang (eds), *China's Regions, Polity, and Economy: A Study of Spatial Transformation in the Post-Reform Era*, Hong Kong: Chinese University Press, 309–350.

Wu, Weiping (2002) "Migrant housing in urban China – Choices and constraints," *Urban Affairs Review*, 38(1): 90–119.

—— (2004) "Sources of migrant housing disadvantage in urban China," *Environment and Planning A*, 36(7): 1285–304.

Wu, Xiaogang and Treiman, Donald J. (2004) "The household registration system and social stratification in China, 1955–1996," *Demography*, 41(2): 363–384.

Wu, Xun and Cao, Yajuan (2002) "*Nongcun shengyu laodongli zhuangyi wenti di zai taolun*" (A review on the shift of the ruarl surplus labor force), *Renkou Yu Jingji (Population and Economics)*, 2002(4): 49–52, 58 (in Chinese).

"*Wuguo shishi baqi fupin gongjian jihua chengji juda*" (The eight–seven poverty alleviation program achieved extensive results) (2001) *Yifan*, [www.yifannet.com/xinwen/guonei/2001-06/20/2220316819014.html], June 19 (accessed June 24, 2004) (in Chinese).

Xiang, Biao (2005) *Transcending Boundaries: Zhejiangcun, The Story of a Migrant Village in Beijing*, Leiden, Netherlands: Brill (translation by Jim Weldon).

Xinhua News Agency (2005) "South China feels acute labor shortage," [www.china.org.cn/english/2005/Mar/121578.htm], March 3 (accessed May 7, 2007).

—— (2006) "No plan to remove migrant workers during Olympics," www.china.org.cn/english/government/181248.htm], September 15 (accessed May 18, 2007).

—— (2007) "Migrant Worker's Suggestions Appear in Premier's Work Report," [www.china.org.cn/english/2007lh-1637.htm], March 6 (accessed April 12, 2007).

Xu, Feng (2000) *Women Migrant Workers in China's Economic Reform*, New York: St Martin's Press.

Xu, Tian Qi and Ye, Zhen Dong (1992) "*Zhejiang wailai nuxing renkou tanxi*" (Analysis of female immigrants in Zhejiang), *Renkou Xuekan (Population Journal)*, 1992(2): 45–8 (in Chinese).

Yan, Shanping (1998) "*Zhongguo jiushi niandai diqu jian renkou qianyi di shitai ji qi jizhi*" (Patterns and mechanisms of interregional migration in China in the 1990s), *Shehuixue Yanjiu (Sociological Research)*, 1998(2): 67–74 (in Chinese).

Yang, Liu (2003) "Rural labor migration choice in China and its impacts on rural households," manuscript.

Yang, Qifan (1991) "*Nannu beijia xianxiang ji qi libi qianxi*" (The phenomenon of southern women marrying to the north and its advantages and disadvantages), *Renkou Xuekan (Population Journal)*, 1991(5): 51–55 (in Chinese).

Yang, Quanhe and Guo, Fei (1996) "Occupational attainment of rural to urban temporary economic migrants in China, 1985–1990," *International Migration Review*, 30(3): 771–787.

Yang, Shangguang and Ding, Jinhong (2005) "*Liudong renkou de chengshi jiuye xiaoying*" (The employment migration and the differential effect of manpower capital in the Yangtze metropolitan delta), *Huadong Shifan Daxue Xuebao (Zhexue Shehui Kexue Ban) (Journal of East China Normal University (Philosophy and Social Sciences))*, 37(3): 82–88 (in Chinese).

Yang, Xiushi (2000a) "Determinants of migration intentions in Hubei province, China: Individual vs. family migration," *Environment and Planning A*, 32(5): 769–787.

—— (2000b) "Interconnections among gender, work, and migration: Evidence from Zhejiang province," in Barbara Entwisle and Gail Henderson (eds), *Re-Drawing Boundaries: Work, Households, and Gender in China*, Berkeley, CA: University of California Press, 197–213.

—— (2006) "Temporary migration and HIV risk behaviors in China," *Environment and Planning A*, 38(8): 1527–1543.

Yang, Xiushi and Guo, Fei (1999) "Gender differences in determinants of temporary labor migration in China: A multilevel analysis," *International Migration Review*, 33(4): 929–953.

Yang, Xiushi, Derlega, V. J., and Luo, H. (2007) "Migration, behaviour change and HIV/STD risks in China," *AIDS Care: Psychological and Socio-Medical Aspects of AIDS/HIV*, 19(2): 282–288.

Yang, Yunyan (1994) *Zhongguo Renkou Qianyi Yu Fanzhan Di Chanqi Zhanlue* (*Long Term Strategies of Population Migration and Development in China*), Wuhan, China: Wuhan chubanshe (in Chinese).

—— (2004) "*Jiushi niandai yilai woguo renkou qianyi de ruogan xin tedian*" (The new characteristics of China's migration since the 1990s), *Nanfang Renkou* (*South China Population*), 19(3): 13–20 (in Chinese).

Yang, Yunyan, Xu, Yangmei, and Xiang, Shujian (2003) "*Jiuye tidai yu laodongli liaodong: yige xindi fenxi kuangjia*" (Employment replacement and labor migration: A new analytical framework), *Jingji Yanjiu* (*Economic Research*), 2003(8): 70–75 (in Chinese).

Ye, Wenzhen (1997) "*Lun shichang jingji dui hunyin guanxi di yingxiang he duice*" (Impacts of the market economy on marital relations, with the countermeasures), *Renkou Yanjiu* (*Population Research*), 21(3): 1–6 (in Chinese).

Yu, Depeng (2002) *Chengxiang Shehui: Cong Geli Zouxiang Kaifang – Zhongguo Huji Zhidu Yu Hujifa Yanjiu* (*City and Countryside Societies: From Segregation to Opening – Research on China's Household Registration System and Laws*), Jinan, China: Shandong renmin chubanshe (Shandong People's Press) (in Chinese).

Yu, Hongwen (2001) "*Cong 2000 nian renkou pucha kan wuguo renkou zhuangkuang di jige tedian*" (Traits of the population situation seen through the 2000 census), *Renkou Yanjiu* (*Population Research*), 25(4): 12–18 (in Chinese).

Yu, Xiong and Day, Lincoln H. (1994) "Demographic characteristics of the migrants," in Lincoln H. Day and Xia Ma (eds), *Migration and Urbanization in China*, Armonk, New York: M. E. Sharpe, 103–128.

Yusuf, Shahid and Nabeshima, Kaoru (2006) *China's Development Priorities*, Washington, DC: World Bank.

Zhai, Jinyun and Ma, Jian (1994) "*Woguo guangdong sheng renkou qianyi wenti tantao*" (Migration in Guangdong Province), *Renkou Yanjiu* (*Population Research*), 18(2): 18–24 (in Chinese).

Zhang, Heather Xiaoquan (1999) "Understanding changes in women's status in the context of the recent rural reform," in Jackie West, Minghua Zhao, Xiangqun Chang and Yuan Cheng (eds), *Women of China: Economic and Social Transformation*, London: Macmillan Press, 45–66.

Zhang, Hong (2007) "China's new rural daughters coming of age: Downsizing the family and firing up cash-earning power in the new economy," *Signs: Journal of Women in Culture and Society*, 32(3): 671–698.

Zhang, Li (2000) "The interplay of gender, space, and work in China's floating population," in Barbara Entwisle and Gail E. Henderson (eds), *Re-Drawing Boundaries: Work, Households, and Gender in China,* Berkeley, CA: University of California Press, 171–196.

—— (2001) *Strangers in the City: Reconfigurations of Space, Power, and Social Networks Within China's Floating Population,* Stanford, CA: Stanford University Press.

Zhang, Li and Zhao, Simon X. B. (1998) "Re-examining China's 'urban' concept and the level of urbanization," *China Quarterly,* 154: 330–381.

Zhang, Ping and Lin, Zi (2000) "Market economic development and reform of the household registration system in China," *Renkou Yu Jingji (Population and Economics),* 2000(6): 35–41.

Zhang, Qingwu (1995) "*Dangqian zhongguo liudong renkou zhuangkuang he duice yanjiu*" (A study on the current situation of population migration in China and its policy implication), in Dang Sheng Ji and Qin Shao (eds), *Zhongguo Renkou Liudong Taishe Yu Guanli* (*The Migration Trend and Management of Population in China*), Beijing, China: Zhongguo renkou chubanshe (China Population Press), 56–64 (in Chinese).

Zhang, Shanyu and Zhang, Maolin (1996) "*Chabie renkou qianyi yu xingbie goucheng diqu chayi de kuodahua*" (Migration between different regions and the enlargement of regional sex composition), *Renkou Xuekan (Population Journal),* 1996(1): 3–10 (in Chinese).

Zhang, Shanyu, Yu, Lu, and Peng, Jizuo (2005) "*Dandai zhongguo nuxing renkou qianyi di fazhan ji qi jiegou tezheng*" (Contemporary female migration in China and its structural characteristics), *Shichang Yu Renkou Fenxi (Market and Demographic Analysis),* 11(2): 13–19 (in Chinese).

Zhao, Shukai (1998) "*1997 nian di nongmin liudong: xin jieduan xin wenti*" (Peasant mobility in 1997: New stages and new problems), manuscript (in Chinese).

Zhao, Yaohui (1999) "Labor migration and earnings differences: The case of rural China," *Economic Development and Cultural Change,* 47(4): 767–782.

—— (2002) "Causes and consequences of return migration: Recent evidence from China," *Journal of Comparative Economics,* 30(2): 376–394.

Zhong, Shuiyang (2000) *Renkou Liudong Yu Shehui Jingji Fazhan* (*Migration and Social Economic Development*), Wuhan: Wuhan University Press (in Chinese).

Zhong, Shuiyang and Gu, Shengzu (2000) "*Dushi fuwuye di fazhan yu liudong renkou di jiuye*" (Development of urban service sector and employment of floating population), *Renkou Yu Jingji (Population and Economics),* 2000(5): 35–38 (in Chinese).

Zhongguo funu (*Chinese Women*) (1997a) "*Dagongmei: bianyuanren di hunlian gushi*" (Working girls: Marriage and romance stories of those living in the periphery), 465: 16.

—— (1997b) "*Quoqiao*" (Magpie Bridge), 460: 58 (in Chinese).

—— (1997c) "*Quoqiao*" (Magpie Bridge), 463: 57 (in Chinese).

Zhongguo qingnianbao (*China Youth News*) (2007) "*Diaocha: jiucheng yishang minzhong renwei youbiyao jinxing huji gaige*" (Survey: more than 90 percent of people think that hukou reform is necessary), [http://news.xinhuanet.com/fortune/2007-02/26/content_5772309.htm], February 26 (accessed May 8, 2007) (in Chinese).

Zhongguo renmin daxue renkou yu fazhan yanjiu zhongxin (Renmin University of China Population and Development Research Center) (2005) "*Zhongguo renkou qianyi liudong yu renkou fenbu yanjiu*" (Research on migration and population distribution in China), in National Bureau of Statistics (ed.), *2000 Nian Renkou Pucha Guojia Zhongdian Keti Yanjiu Baogao* (*National Level Key Research Reports on the 2000 Census*), Beijing: China Statistics Press, 912–1035 (in Chinese).

Zhou, Daming (1992) "*Zhujiang sanjiaozhu wailai laodong renkou fenbu tezheng ji yidong quxi fenxi*" (The nonnative labourers of the Pearl River Delta: An analysis of its distribution characteristics and migration tendencies), in Zhongshan University Research Centre of Pearl River Delta Economic Development and Management (ed.), *Zhujiang sanjiaozhu jingji fazhan huigu yu qianzhan* (*Economic Development of the Pearl River Delta: A Retrospect and Prospects*), Guangzhou: Zhongshan University, 271–280 (in Chinese).

—— (1998) "*Zhongguo nongmingong di liudong: nongmingong shurudi yu shuchudi bijiao*" (Migration of peasants in China: Comparison of origins and destinations), in Zhongshan University Anthropology Department (ed.), *Waichu Wugong Diaocha Yanjiu Wenji* (*Collection of Papers on Surveys of Peasant Out-Migration*), Guangzhou, China: Zhongshan University Anthropology Department, 3–23 (in Chinese).

Zhou, Hao (2002) "A review, summary and discussion of population migration study in China," *Renkou Yu Jingji* (*Population and Economics*), 2002 (1): 56–59 (in Chinese).

—— (2004) "*Zhongguo renkou qianyi de jia ting hua qushi ji yingxiang yinsu fenxi*" (The analysis on the trend and factors of family migration), *Renkou yanjiu* (*Population Research*), 28(6): 60–7 (in Chinese).

Zhou, Yixing and Ma, Laurence J. C. (2003) "China's urbanization levels: Reconstructing a baseline from the fifth population census," *China Quarterly*, 173: 176–196.

—— (2005) "China's urban population statistics: A critical evaluation," *Eurasian Geography and Economics*, 46(4): 272–289.

Zhu, Yu (2003) "The floating population's household strategies and the role of migration in China's regional development and integration," *Internationl Journal of Population Geography*, 9: 485–502.

—— (2007) "China's floating population and their settlement intention in the cities: Beyond the *Hukou* reform," *Habitat International*, 31(1): 65–76.

Index

administrative hierarchy and units 23–7, 31, 174n13, 175n24
age: discrimination 105–7; marriageable 86, 140, 143, 178n61
agriculture: and circular migration 93, 164; commercialization of 149; decollectivization of 10; dislike of 6, 72, 93, 121–2, 131, 163; effect of WTO on 163; feminization of 90; labor force in 6, 148, 167; input to 94, 120; and marriage migration 148–50; migration's effect on 18, 120–4, 135; as an occupation category 100–2, 145, 147; planting and harvesting seasons of 93; policy for 4; productivity of 6, 18, 120–1, 135, 149, 156; as a source of livelihood 6, 93, 122, 163; subsidies for 118; transfer of value from 4, 40, 44; women's contribution to 9
Algeria 46
Anhui 29, 31–2, 35–6, 38, 51, 121, 124, 143, 151, 178n62; see also Sichuan and Anhui Household Surveys; Sichuan and Anhui Interview Records

bargaining-power hypothesis 137
Beijing 10, 29, 31–2, 34, 38, 41, 51, 89, 97–9, 114–15, 127, 132–3, 135, 144, 150, 168–9, 171–2, 175n22; hukou in 51, 175n27, 175n30, 176n31, 178n60, 178n70; the Olympics in 162, 178n69
bendiren (natives, local people) 111
best of both worlds 12–13, 17, 123, 165–6, 171
"blind flows" (mangliu) and drifters 44, 46, 97
bottom-up approach 8, 11, 13, 53
brain drain 122
brain-drain reversal 122, 125
brain gain 122
breadwinner 80, 85, 90, 129
bride-price 140, 142, 158
"bringing the state back in" 2

Cantonese (baihua) 155
Central Asian Republics 31, 37
central planning 3, 17, 42–3, 45, 48–9
chain or snowball migration 142, 155, 161
children: education for 72, 88, 92, 114, 118, 126–8, 171; and hukou 87, 171, 175; left-behind 87, 89–92, 94, 122, 129–31, 133, 160; migrating with parents 57, 90, 127, 164, 166, 171; and the one-child policy 177; school-age 77, 92, 127
"China's rise" 162
Chinese Communist Party (CCP) 3, 170, 176n32; membership of 10, 42, 14
Chinese socialism 52
Chongqing 30, 49
circular migration (circulation, circularity) 12–13, 18, 164–8, 172; of husbands 17; as a long-term practice 164; and return migration 124; and seasonal migration 165; and "wild geese households" (yan hu) 164
citizenship: dual 12; urban 47, 53, 172
city size: control of 48; and hukou 51–2
collectivization 3, 6, 43; and communal production 5; and communal protection 7; and decollectivization 10, 141
"common prosperity" 170
Confucian ideology 9, 139
Constitution, China's 43, 46
construction work and construction workers 5, 89, 91, 93, 97, 100, 103–4, 107–8, 113, 162, 165, 169–70, 176
cost-of-living differentials 165

dagong 7, 111, 155, 166; marriage 144, 159–61; school 127; *see also* migrant labor regime; migrant work; migrant workers; migrants
dagongmei (working women, working sisters) 7, 84, 111; *see also* maiden workers; women migrants
dagongzai (working men) 7, 111
data 13–16; and anonymity 13; census 13–14, 17, 19–39; multiple sources of 13; qualitative 13; quantitative 13
debt: as a migration reason 72, 91; and use of remittance 94, 118–19
decollectivization *see* collectivization
Deng Xiaoping 3
development: disparity 37, 143, 151, 161; and hukou 40; level of 31, 161; and migration 37–9, 48; of rural areas 2, 6, 117, 121, 123, 125–6, 172; strategy 4–5, 44, 47, 52–3; urban 5, 48, 52–3, 112
developmentalist state 3–6, 10, 40, 53, 116
disparity *see* inequality
distance: as a factor of migration 36–8, 72, 147; friction of 12, 17, 21, 37–9, 71, 76; of migration 17, 21, 61, 64, 67, 72, 75–6, 80, 82, 94, 139, 141–2, 144, 146–7, 149–52, 154–5, 161, 175, 176n34, 176n38, 176n40
division of labor 8, 11, 17, 93; gender 9, 75, 85, 89–92, 94, 129–31, 141, 144, 149, 160–1; reverse 91–2, 131; intergenerational 90, 92
divorce *see* marital disruption
domestic work (and nanny) 5, 98–9, 104–6, 112, 114–15, 133
dowry 140

education *see* children; hukou; migrants; migration reason; occupations; women
employment: allocated by the state 17, 45–6, 62; channels of 73, 81, 97–9, 103; as contract workers (*hetonggong*) 108; as a factor of migration 39, 43, 54, 71, 113, 122, 138, 163, 167; as formal workers 175n21; generated by migrants 112, 125; migration's pressure on 49; negative growth in 35; in non-farm work 10, 126, 163–4; practices 17, 95, 116; as "regular employees" (*zhengshi zhigong*) 108; as temporary workers (*linshigong*) 108; for wage 176n39
enterprises: collective 46; foreign-invested 103; private-owned 103, 115, 176n31; state-owned (SOE) 42, 46, 108; township-village (TVE) or rural industrial 163, 107, 178n65

family: concept of 8; extended 8, 128; *see also* household
farmland: abandonment 121; access to 7, 22, 41, 57, 93, 171; allocation 6, 148; carrying capacity of 123; in Guangxi 156; lack of 71–2, 85; leasing or subcontracting of 7, 90, 121; and population registration 43; as security and insurance 94, 123, 165; taking care of 7, 89–93, 128
female migrants *see* women migrants
feminism, Maoist version of 10
feminist: approach 8; awareness 10; methodologies 8
Five-Year Plan 47, 49; Eleventh 52, 170–2; Seventh 28; Tenth 52
floating population 1, 13, 19, 23–4, 163, 173n12; and household strategies 88; intercounty 23–4, 39; intracounty 23–4; and *liudong renkou* 24, 111; and *wailai renkou* 24, 111; *see also* hukou; migrants; migration
Fujian 12, *29*, 31, 34, 38, 93, 107, 150, 167

Gansu *30*, 51, 120–1
Gaozhou 14–15, 90, 106, 130–1, 133, 152–60
GDP 28–31, 122
gender 9–11, 17, 75–94; and Confucianism 9–11; constructions of 92; differentials in migration 17, 65–6, 75–84; equality 9; identity 10–11; inequality 86; and the labor market 5, 104–7; and migration distance 67, 76–7, 80; and migration propensity 76–9; and migration reason 59, 80–3; roles and relations 2, 3, 8–11, 15, 17–18, 75, 81, 83–94, 128–2, 136–9; and social networks 99, 114–15; *see also* division of labor; patriarchy
grand international cycle theory 3
grandparents 92, 160
gravity model 38
Guangdong 14–15, *29*, 31–2, 34, 36–9, 85, 89, 93, 104–7, 114, 120, 124, 131, 133, 150, 152–5, 157, 159–60, 167, 169
Guangxi *29*, 34–6, 38, 150–1, 154–8
Guangzhou 14, 51, 69–72, 84, 96–8, 103–4, 111, 113, 174, 176n38
Guizhou *30*, 31, 35–6, 143, 150–1

Hainan *29*, 34, 36, 159

Hebei 29, 32, 34, 38, 51, 124, 150
Heilongjiang 30, 31, 36–7, 151
Henan 6, 30, 31–2, 37, 120, 125, 154, 171, 177
house building and renovation: in Gaozhou 156; as a migration reason 72, 120; as a reason to stay in the village 165; and use of remittance 18, 94, 118, 131, 134, 160
household: and hukou 41, 43; and the migrant 11, 118–21, 135; and the state 2, 5–6; two-generation 68, 160; as a unit of analysis 7, 11, 53, 83, 93, 135; *see also* family; household strategies; split household
household division of labor *see* division of labor; split household strategy
household formation 144
household income function 120
household registration *see* hukou
household responsibility system 6
household strategies: approach 1, 6–9, 11–13, 75–94, 165; collective and negotiated 93, 128, 165; and risk 89, 141; and social relations and roles 15, 93, 137–41; *see also* division of labor; split household strategy
"housing change" as a migration reason 54–5, 61–2, 80, 82, 176n34
Hu Jintao 18, 168, 170, 172
Hubei 30, 34, 36, 51, 124
hukou (household registration): abolition of 88, 171; agricultural (*nongcun*) 41–3, 174n20; as an ascribed status 42, 53; being commodified 49–51, 175n24; "blue stamp" 50–1, 175n25; and children's education 114, 171; and city size 43, 50–1, 171, 175n24, 175n28; conversion from agricultural to non-agricultural (*nongzhuangfei*) 42; criticisms of 47–9; and dualism 1, 48, 50, 166, 172, 176n35; and the floating population 23–4; and Five-Year Plans 52, 171; inherited from parents 86–7, 175n26, 175n27; and the labor market 95–6, 101, 104, 106–8, 116, 175n23; location (*suozaidi*) 17, 20–3, 40–3; management of 48; and marketization 48; and migration 22, 45–6, 55, 58–69, 79, 88, 93–4, 142, 146, 148, 164; non-agricultural (*feinong*), 41–3, 174n20; paradigm 2, 11–13, 18, 171; and population registration 43, 46–9, 52; reforms of 5, 12, 16–17, 47, 49–53, 162, 171, 175n26, 175n29, 176n31; regulations of 40, 44, 47; and *renhu fenli* 22; rural 43, 48, 93–4; "self-supplied food grain" 50; and the state 4–5, 40, 43–7, 116; system 1, 4–5, 17, 21, 23, 39–53; and "temporary residence permit" 50; type (classification, *leibie*) 40–3, 51; and university education 57, 66, 74; urban 4, 11–13, 22, 43, 48, 50, 57, 93–4, 142, 148, 164, 171, 175n26; and welfare 40, 46–7
human capital: as a factor of migration 2, 47, 59, 65, 69, 74, 79; and hukou 17; impacts of migration on 122, 125; and the labor market 95; and marriage migrants 149, 161
Hunan 30, 31–2, 35–6, 38, 49, 51, 120, 124, 151, 154, 157, 159
hypergamy 137, 139; *see also* spatial hypergamy

identity card 50, 106, 110
income (and monetary return): as a criterion for hukou 51; in Gaozhou 154; gender gap in 85; and job search 96, 103; as a marriage attribute 140, 143, 178n60, 178n64; from migrant work 8, 11, 93, 104, 118–21, 123, 126, 163, 177n47, 177n49, 178n55; as a migration reason 2, 6, 7, 17, 70–2, 74, 88–9, 109, 116, 123, 125, 138, 167; rural 18
industrialization: export-oriented and labor-intensive 3–4, 12; migrants' role in 101; on the cheap 4, 44–5, 52
inequality (disparity, uneven development): due to personal attributes 67; policies to address 168, 170; regional 1, 28, 31, 37, 123, 161; rural–urban 11, 44, 48, 53, 117, 123, 135, 177n46; *see also* gender
inside–outside: ideology 9, 17, 87–90, 94, 129–31, 135; perception 87, 132, 177n42
institutional: approach and explanations 2, 11, 65, 67, 97, 116, 139, 149; divide and status 4–5, 11, 17, 40–1, 43–5, 52–3, 69, 72, 74, 88, 95, 110, 127, 135, 149, 161, 176n33; legacy 50
institutions: socialist 3, 5, 47, 50, 55; state 3, 6
international migration (immigration) 1–2, 12, 172, 173n1; Canada 50; Germany 48; Japan 48; Mexico–US 7; policy on 50; US 48, 172

"iron rice bowl" 45

Jiang Zemin 170
Jiangsu *29*, 31–2, 34, 36, 38, 73, 103–4, 108–10, 121, 124–5, 142–3, 150, 152, 154, 178n64
Jiangxi *29*, 31, 35–6, 38, 118, 124, 151
Jilin *30*, 31, 36–7, 120
job mobility 5, 96, 103, 113–15; and labor allocation 4, 45, 104
job search 45, 70, 96–100, 103; and advertisement 96; and employment agencies 96–100; and recruitment 96–100; and social networks 74, 96–100, 115

labor market: formal and informal segments of 103; and gender 5, 10, 18, 84–5, 90, 104, 107, 139, 161; and gender wage gap 137; in Guangzhou 14; and hukou 11, 48, 95–109; information about 74, 96–7, 168; international 12; and ownership sectors 96, 103; practices 105–9; in the pre-reform period 45, 104; primary sector of 5; returns 104–9; segmentation 5, 17, 48, 95–109; and social networks 114–15; *see also* migrant labor regime; occupations
labor migrants *see* migrant work; migrant workers
labor migration *see* migrant work; migrant workers
labor protests and resistance 18, 167–70
labor shortage 18, 112–13, 162–3, 166–8
labor surplus: agricultural or rural 4, 6–7, 44, 85, 121–2, 135, 163, 167, 177n54; as a migration reason 7, 71–2, 112, 163; in Sichuan and Anhui 15
Li Peng 170
Liaoning *29*, 34, 36
"lychee houses" 156

maiden workers 79, 84–5, 88, 94; and stereotypes 105, 107; *see also dagongmei*
marital disruption 131–3, 143, 178n62; and extramarital affairs 133; and geographical separation between spouses 133
marriage: age of 9, 86, 88, 140, 143, 157, 178n61; among kin 139; arranged 10, 86, 140; and attribute (*tiaojian*) matching and tradeoff 17, 19, 140–2, 144, 149–50, 156–9; broker and introduction center for 142–3; and class origins 140; *dagong* 144; expenses as a migration reason 72, 118; and family or household 8, 178n68; gender roles within 8–9, 88–9, 128–31, 135, 137; as an institution 92, 133; and the labor market 105, 141; as a life event 137, 139; matchmaker (*meiren, jieshao ren*, intermediary) 140, 154–5; and migration 14, 18, 77, 85–9, 94, 129–30, 135, 137–41; and pragmatism 138; rate of 79; squeeze ("wife shortage") 138; transactional and pragmatic 139–40; *see also* marriage migration; patrilocal tradition and exogamy
Marriage Law 9–10, 140
marriage market 49, 86, 158; and labor market 141; of peasant women 139; urban 142, 144, 149
marriage migration 14, 18, 41, 55–65, 70, 79–82, 101, 137–61, 178n59, 178n62; to Belgium 139; compared to labor migration 143–50; *dagong* 159–60; and distance 139, 147; to Gaozhou 153–60; to Germany 138; and hukou 79; in India 138; international 138–9; to Japan 138; and labor migration 143–4; and location 138, 141–3; long-distance 139–54; and "mate-finding problems" 159; in Puerto Rico 139; short-distance 141, 154; spatial patterns of 150–2; and wedding expenses 158; *see also* patrilocal tradition and exogamy; spatial hypergamy
"matching doors" (*mengdang hudui*) 140; *see also* marriage
migrant children schools *see* children
migrant community 114–16, 172
migrant labor regime 2, 3–5, 53, 97, 99; and benefits 105, 108; and discipline 109, 168; and discrimination 105–7; and exploitation 107–9, 164; and living spaces 109–11; and work disruption 105, 107, 109
migrant stock 38
Migrant Women's Club 168
migrant work: as a long-term strategy 87, 89, 93, 113, 115; marginal return to 120; opportunity cost of 120–1; as a way of life 18, 122, 163; *see also* migrant labor regime; migrant workers; migrants; migration
migrant workers: as competition to urban workers 112–13; labor rights of 5, 171; recruitment of 45, 72–3, 95–9, 105, 116; resistance of 167–70;

migrant workers *continued*
  see also migrant labor regime; migrant work; migrants; migration
migrants: agency of 8, 13, 172; choices of 13, 106, 108, 149, 168; consumption of 112; and crime 49, 113, 175; destinations of 12, 14, 24–7, 29–39, 58, 60–1, 68–81, 97, 99, 113, 118, 142, 146, 148–53, 161; educational attainment of 60–1, 64, 67–9, 79, 101, 147, 158, 175; elite 50, 61, 95; experiences of 13, 16, 95–116; as a hybrid class 11, 17, 123, 166, 171; identity of 166; impacts on inequality 123–4; impacts on lifestyle 120, 133–6; impacts on rural areas 117–36; impacts on urban areas 112–13; intention of 12–13, 88–9, 94, 113–14, 116; as "lateral" movers 71; and manual work 91, 112; marital status of 58, 60, 68, 76, 145; and menial work 100, 103; origins of 2, 11–12, 24–7, 29–39, 60–1, 64–80, 88, 97, 111, 116–17, 143, 145, 148, 150–5, 157, 161, 177n49; as outsiders 17, 48, 53, 95, 110–13, 116, 127, 144; permanent 21–4, 42, 56, 60–2, 64–80, 95–7, 101, 102–5, 116, 148; permits 40, 47, 50, 98, 110–11, 175n30; productivity of 112, 120; replacement 91; responses of 113–16, 163–70; second-generation 16, 112, 163–4, 168; selectivity of 69, 74, 76, 79, 113, 122; sex ratio of 57–8, 60, 64, 67–8, 76; social exclusion of 111–12, 127; social impacts of 128–36; social protection for 170–2; temporary 4–5, 11–12, 14, 21–4, 39, 41, 43, 49–50, 53, 56, 60–80, 84, 95–7, 101, 102–5, 110, 116, 148; as "upward" movers 42, 71, 139; and urban infrastructure 49, 51, 113; see also migrant labor regime; migrant work; migrant workers; migration
migration: control of 4, 19, 42–9, 52, 113; counterstreams of 37–8; de facto 22; de jure 22; of the entire family 164; fields 36–8; flows 17, 32–9, 150–1; formal 22; hukou 22; impacts on sex ratio 158; informal 22; and *liuqian* 24; net 28–9, 34–5, 150–2; non-hukou 22; non-plan 22; plan 22; and poverty alleviation 8, 43, 117–19, 123, 131, 135, 138–9; and *qianyi* 24; and *qianyi renkou* 24; redistributing population 31, 38; rural–rural 24–7, 57–61, 145–6; rural–urban 1, 4, 9, 11, 17, 19, 22, 24–7, 40, 43–7, 57–61, 117, 125, 145–6, 174n14; self-initiated versus. state-sponsored 17, 22–3, 39, 55–67, 74, 176n35; spatial (and regional) patterns of 20–39; streams of 37–9, 69, 74, 97, 142, 152, 161; theories on 2–13; two-track 1, 17, 22, 61, 64, 67, 69, 74, 95, 172; urban–rural 24–7, 57–61; urban–urban 24–7, 57–61; volume (magnitude) of 17, 19–39, 163; see also migrants
migration measure 19; flow 17, 19–24, 39; stock 17, 19, 23–4, 39
migration rate 19, 21, 23–4, 34–5, 76–8, 118; age-specific 78
migration reason 54–74; economic 54–63, 70, 74, 80–3, 162–3; education fee 72; family 70, 80–3, 127; friends/relatives 55–63, 80–3; housing change 54–63, 80–3, 176n34; industry/business 55–65, 80–3, 144–7; job assignment 55–63, 80–3; job-related 70–1, 80–3, 144–7; job transfer 55–65, 80–3; joining family 55–63, 80–3; life-cycle 54–63, 71, 74, 85, 147; retirement 54–63, 80–3; self-improvement 71–2; social 54–63, 71, 74, 80–3; study/training 55–63, 67, 80–3, 101; see also agriculture; debt; farmland; house building and renovation; income; labor surplus; marriage; marriage migration
Ministry of Agriculture 15–16, 46, 124
Ministry of Public Security 43, 50, 52, 84, 176n35

Nanjing 51, 73, 103, 112, 115, 174
narratives 13, 15–16, 97, 103, 128, 135, 166
National People's Congress (NPC) 40, 170
networks: gender-based 99, 114–15; and marriage migration 152, 155; of the natal family 139–40; native-place 17, 114–16; social 17–18, 57, 71–2, 74, 97–100, 114–16; *tongxiang* 2, 38–9, 69; see also job search; labor market; marriage migration; migrant stock; migration reason
New Economics of Migration theory (NEM) 11
Newly Industrializing Economies (NIEs) 3–4
Ningxia *30*, 34, 171
nongmingong see dagong; migrant workers

occupations: definitions of 100; and education 101; and gender 115, 131, 137, 140; and hukou 41, 43, 45, 101–4; and the labor market 100–4, 115, 175; and marriage migration 101, 145–6, 149–50
Olympics, the 162, 178n69
one-child policy 10, 177n41

patriarchy 9–10, 17, 85, 88, 94, 129, 137; see also gender
patrilocal tradition and exogamy 9, 139
Pearl River Delta 113, 153, 160, 166–7, 169
peasants (*nongmin*): dream of 120; identity of 144, 166, 171–2; problem of 122; and the state 3–7, 40, 44–7, 52–3, 176n32
permanent migrant paradigm 2, 11–13, 18, 76, 171
population size 29–32, 36, 38, 49, 122
poverty 7–8, 43, 46, 117–19, 123, 131, 135, 138–9, 170, 175n22, 178n56, 178n71
power relations 8, 11, 15, 88, 137, 171–2; see also social hierarchy
prices of agricultural goods 4, 6, 44–6, 52; and "scissors gap" 45–6; see also transfer of value from agriculture to industry
propiska 47
public security authorities 42–3, 46, 50, 52, 84, 110, 112, 170

Qinghai *30*, 34

Ravenstein, E.G. 75
remittances 8, 11, 18, 93–4, 117–23, 135, 160, 163, 165, 167, 177n50, 177n51, 177n52; and gender 9, 130–1, 133; use of 18, 94, 118–20, 135
return migrants: as entrepreneurs 16, 125–6; family 127; selectivity of 126; success versus failure 124–7
return migration 4, 12–13, 16, 19–20, 56, 64, 124–8, 135, 163, 165–6, 168, 171–2; and age, health and sickness 93, 107, 128; and children 92, 114, 127, 163; definitions and estimates of 124; and discrimination 114; and farmland 94; impacts of 18, 125–6, 135; intention of 12, 88–9, 94, 105, 124, 164; and marriage 86–8, 92, 94, 129, 131, 144; and rustication 37, 42; and split households 8, 11–12, 17–18, 89, 160

Russia (the former Soviet Union) 3–4, 43–4, 47

sannong problem, the 122
self-employment 96, 103
Shandong *29*, 32, 34, 36, 38, 150
Shanghai *29*, 31–2, 34, 36, 38, 50–1, 85, 99, 107, 112, 114, 126–9, 162–3, 167, 175n25, 178n70
Shanxi *29*, 34, 49
Shenzhen 7, 49–50, 84, 104–5, 109–10, 113, 128, 153, 159, 163, 169, 175n25
Shijiazhuang 12, 51
Sichuan 6, 15, *30*, 31, 35–6, 38, 120–1, 124, 144, 150–1, 168, 177n52; see also Sichuan and Anhui Household Surveys; Sichuan and Anhui Interview Records
Sichuan and Anhui Household Surveys 15–16, 24, 84, 118, 124, 126–7
Sichuan and Anhui Interview Records 7, 9, 15–16, 24, 31, 35, 69, 71–4, 84–93, 95, 96, 103, 105, 107, 118–20, 134, 148, 162–7, 169–70
social hierarchy: based on gender 15, 18, 86; based on geographic origin 48; see also power relations
"socialist harmonious society" 170
socialist market economy 3
socialist transition 3, 47
sojourners 89
South Africa 46, 48, 177n48
spatial hypergamy 131, 137, 141–4, 154–60; see also hypergamy; marriage; marriage migration
split household strategy 8, 11, 17, 85, 87–94, 114, 122, 130, 133, 135, 144, 171, 160–1; see also division of labor; household strategies
Spring Festival, the 15, 73, 92–3, 98–9, 109, 115, 134, 164–5; and cost of transportation 164
"spring water flowing east" 152; see also marriage migration
State Council, the 43, 50–1, 87, 125, 171, 175, 179n71

taxes (fees): agricultural 6; due to migration 43, 50, 98–9, 110, 127, 175n24
"three economic belts" 28
Three Gorges Project 176n34
Tianjin *29*, 31–2, 34, 109, 115, 150
tongxiang (*laoxiang*, people from the same native place, fellow villagers): as intermediary 140, 154–6, 161;

tongxiang continued
  as migration company 72–3; as a
  migration reason 72–3; and social
  networks 97–100, 108, 114–16, 177n45
township-village enterprises (TVE) *see*
  enterprises
transfer of value from agriculture to
  industry 40, 45–6, 52; *see also* prices of
  agricultural goods

unemployment (underemployment) 6, 46,
  104, 113
"unified purchase and marketing" (*tonggou tongxiao*) 45
"unified state assignment" (*tongyi fenpei*) 4, 45
United Nations Fourth World Conference
  on Women 10
urbanization 1, 12, 19, 24–5, 49, 52, 163,
  173n2, 174n13
US census 19–20

Vietnam 12, 46, 178n67

wailainu (*wailaimei*, women from outside) 142
wailairen (*waidiren, wailai renkou*, people
  from outside) 24, 111
Wen Jiabao 18, 168, 170, 172
Western Development ("Go West")
  program 170
women: agency of 18, 131, 137–9, 141,
  161; career prospect of 137; de-skilling
  of 90, 131; independence of 128–9, 133,
  135; labor force participation of 7, 147;
  life cycle of 85–8, 129–30, 147, 178n61;
  self-image of 128, 135; status of 9–10,
  76–80, 85–6, 128, 131–3, 139, 142, 144;
  well-being (*xinfu*) of 139; work loads of
  130–1; *see also* division of labor;
  gender; inside–outside; marriage;
  patriarchy; social hierarchy; women
  migrants
women migrants 17, 71–3, 75–88, 94, 134,
  141–61; educational attainment of 10,
  76–9, 144–9; husbands of 149–50; as
  replacement migrants 91; as tied movers
  (passive movers, trailing spouses) 60,
  137; and wage work 128–9, 176n39; *see
  also dagongmei*; maiden workers;
  marriage migration; women
World Expo, the 162
World Trade Organization (WTO), the
  162–3

xiagang (laid-off) workers 113, 165
Xian (Xi'an) 51
xiaokang 170, 179n71
Xinjiang *30*, 31, 33–4, 36–8

Yangtze River Valley 167
Yunnan *30*, 33–4, 36, 38, 150–1, 154

Zhejiang *29*, 31, 36, 38, 120, 124, 142, 150
Zhejiang Village 115–16, 172, 175
Zhengzhou 6, 51, 73
Zhou Dynasty 43, 46
Zhuhai 51, 175n22, 175n29